AF595029

Shallow Water Equations in Hydraulics: Modeling, Numerics and Applications

Editors

Anargiros I. Delis
Ioannis K. Nikolos

MDPI • Basel • Beijing • Wuhan • Barcelona • Belgrade • Manchester • Tokyo • Cluj • Tianjin

Editors
Anargiros I. Delis
Technical University of Crete
Greece

Ioannis K. Nikolos
Technical University of Crete
Greece

Editorial Office
MDPI
St. Alban-Anlage 66
4052 Basel, Switzerland

This is a reprint of articles from the Special Issue published online in the open access journal *Water* (ISSN 2073-4441) (available at: https://www.mdpi.com/journal/water/special_issues/Shallow_Water_Hydraulics).

For citation purposes, cite each article independently as indicated on the article page online and as indicated below:

LastName, A.A.; LastName, B.B.; LastName, C.C. Article Title. *Journal Name* **Year**, *Volume Number*, Page Range.

ISBN 978-3-0365-3317-9 (Hbk)
ISBN 978-3-0365-3318-6 (PDF)

Contents

About the Editors

Anargiros I. Delis is a Mathematician, Associate Professor of "Computational Mathematics" at the School of Production Engineering & Management of the Technical University of Crete, Greece, where he has been working since 2003. He is a graduate of the Department of Mathematics of the University of Crete (1993); a holder of an M.Sc. in Numerical Analysis & Computing from the University of Manchester Institute of Science & Technology (UMIST), U.K.; and received his PhD from the University of the West of England, Bristol, U.K., in Applied and Computational Mathematics in 1998. His research focuses on the areas of computational hydrodynamics of free surface flows, computational fluid dynamics, and more generally in the area of scientific computing.

Ioannis K. Nikolos is a full professor with the School of Production Engineering & Management, Technical University of Crete, Greece, and Director of the Turbomachines & Fluid Dynamics Laboratory (TurboLab: TUC). Prof. Nikolos has obtained his Diploma (1990) from the School of Mechanical Engineering, National Technical University of Athens (N.T.U.A.), and a Ph.D. degree (1996) from the same school in Turbomachines (Lab. of Thermal Turbomachines, N.T.U.A.). He joined Technical University of Crete in 1998. Prof. Nikolos has more than 30 years of experience in R&D projects funded by the EU, industry, and the Greek State. His research work is in the fields of computational fluid dynamics (CFD), computational heat transfer, computational engineering, turbomachines, and engineering design optimization using AI (artificial intelligence) tools.

Editorial

Shallow Water Equations in Hydraulics: Modeling, Numerics and Applications

Anargiros I. Delis * and Ioannis K. Nikolos

School of Production Engineering & Management, Technical University of Crete, University Campus, 73100 Chania, Greece; jnikolo@dpem.tuc.gr
* Correspondence: adelis@science.tuc.gr

Abstract: This Special Issue aimed to provide a forum for the latest advances in hydraulic modeling based on the use of non-linear shallow water equations (NSWEs) and closely related models, as well for their novel applications in practical engineering. NSWEs play a critical role in the modeling and simulation of free surface flows in rivers and coastal areas and can predict tides, storm surge levels and coastline changes from hurricanes and ocean currents. NSWEs also arise in atmospheric flows, debris flows, internal flows and certain hydraulic structures such as open channels and reservoirs. Due to the important scientific value of NSWEs, research on effective and accurate numerical methods for their solutions has attracted great attention in the past two decades. Therefore, in this Special issue, original contributions in the following areas, though not exclusively, have been considered: new conceptual models and applications; flood inundation and routing; open channel flows; irrigation and drainage modeling; numerical simulation in hydraulics; novel numerical methods for shallow water equations and extended models; case studies; and high-performance computing.

Keywords: shallow water equations; free surface flows; modeling; hydraulic engineering; computational methods; simulation

Citation: Delis, A.I.; Nikolos, I.K. Shallow Water Equations in Hydraulics: Modeling, Numerics and Applications. *Water* **2021**, *13*, 3598. https://doi.org/10.3390/w13243598

Academic Editor: Helena M. Ramos

Received: 15 November 2021
Accepted: 14 December 2021
Published: 15 December 2021

1. Introduction

In hydraulic engineering, the modeling and simulation of free surface water flows plays an important role in many real-life practical applications. An important feature of free surface water flows is that they are unbound in space, the limits of the spatial domain being an unknown of the problem to be solved. Problems in which the limits of the fluid are unknown and unsteady include, among others, dam break and flooding flows, tidal flows in coastal water regions, nearshore wave propagation with complex bathymetry structure, tsunami wave propagation and ocean modeling. The full flow field in most such applications can be described by Navier–Stokes equations. However, qualitative and/or quantitative approximations of the actual solution are given by approaches based on simplified equations. This is done in a systematic effort usually to overcome the need for excessively demanding numerical techniques to resolve Navier–Stokes equations (possibly supplied with appropriate turbulence closure models). A widely used approach is that of the 2D depth-averaged models [1,2]. The 2D character of the free surface flow is usually enforced by a horizontal length scale which is much larger than the vertical one, and by a velocity field which is quasi-homogeneous over the water depth. This small ratio between the vertical and horizontal length scales characterizes many engineering applications, mainly in river and coastal engineering. As such, many geophysical flows can be modeled by the well-known non-linear shallow water equations (NSWEs), also called Saint-Venant (SVEs) equations, and closely related models. From a mathematical point of view, these models constitute a system of partial differential equations, mainly of the hyperbolic type, also referred to as hyperbolic systems of balance laws.

Despite the relative simplicity of NSWEs, the shallow water assumption is valid in many applications in hydraulics and as such has a long tradition of providing a scientific basis for engineering practice. To this end, shallow water equations arise in modeling water flows in rivers, canals, lakes, reservoirs, coastal and urban areas and many other situations in which the water depth is much smaller than the horizontal length scale of motion. Hence, shallow water and closely related models are widely used in oceanography and atmospheric sciences to model, among others, flood waves and hazardous phenomena such as hurricanes/typhoons and tsunamis. In recent years, new conceptual models and applications based on NSWEs have emerged. These include, among others, models for flood inundation and routing, sediment transport and morphodynamic models, avalanches, pollutant transport in water models and dispersive wave propagation ones. Furthermore, in recent decades, many efforts have been devoted to develop one- and two-dimensional numerical models for unsteady free surface flows, with the most widely used mathematical framework to be the one based on the 2D NSWEs [3]. The need for accurate and efficient numerical methods has resulted in an ongoing quest for such methods that are of higher-order spatial and temporal accuracy that can effectively predict the most relevant physical phenomena.

In this Special Issue, we attempted to include manuscripts related to recent advances in terms of modeling, numerical methods and applications closely related to the implementation of the NSWEs.

2. Summary of This Special Issue

In [4], the authors presented an extensive numerical investigation for studying the accuracy and efficiency of three common shallow water finite volume solvers, namely the HLLC, Roe, and Central-Upwind (CU) schemes. Four cases dealing with shock waves and the wet–dry phenomenon were selected. All schemes were provided in an in-house code NUFSAW2D, the model of which was of second-order accuracy in space wherever the regimes were smooth and robust when dealing with strong shock waves—and of fourth-order accuracy in time. To give a fair comparison, all source terms of the 2D NSWEs were treated similarly for all schemes, namely the bed-slope terms were separately computed from the convective fluxes using a Riemann-solver-free scheme—and the friction terms were semi-implicitly computed within the framework of the RKFO method. Two important findings were shown by the presented simulations. Firstly, highly efficient vectorization could be applied to the three solvers on all hardware used. This was achieved by guided vectorization, where a cell-edge reordering strategy was employed to ease the vectorization implementations and support the aligned memory access patterns. Secondly, it was shown that for the four cases simulated, strong agreements by all schemes were obtained between the numerical results and observed data, where there were no significant differences. However, in terms of efficiency, the CU scheme was able to outperform the HLLC and Roe schemes with average factors of 1.4× and 1.25×, respectively. Finally, it was demonstrated that the edge-driven level, especially the reconstruction technique and solver computations, were the most time-consuming parts, which required 65–75% of the entire simulation time. This shows that some more "aggressive" optimization techniques will still constitute a hot topic for future studies to make shallow water simulations more efficient, particularly in the edge-driven level. Since simulating shallow water flows—especially complex phenomena that require performing long real-time computations as part of disaster planning such as dam-break or tsunami cases—on modern hardware nowadays, and even in the future, is becoming increasingly common, focusing simulations only on numerical accuracy but ignoring the performance efficiency may not be an option anymore. Wasting the performance is obviously undesirable the excessive time required by such long real-time simulations. Modern hardware offers many features for gaining efficiency, one of which is vectorization, which can be regarded as the "easiest" way for benefiting from the vector-level parallelism, and is thus non-trivial.

The authors in [5] aimed to investigate the dynamic process of the transformation between different channel patterns with different control variables. They proposed an extended 2D depth-averaged numerical model which incorporates the hydrodynamic, sediment transport and a river morphological adjustment sub-model. The model was solved in an orthogonal curvilinear grid system by using the Beam and Warming alternating-direction implicit (ADI) scheme. The sediment transport sub-model includes the influence of non-uniform sediment with bed surface armoring and a correction for the direction of bed-load transport due to secondary flow and transverse bed slope. The bank erosion sub-model incorporates a simple simulation method for updating bank geometry during either degradational or aggradational bed evolution. Then, the extended 2D model was applied to duplicate the evolution of the channel pattern with variations in flow discharge, sediment supply and bank vegetation. Complex interactions among the flow discharge, sediment supply and bank vegetation leads to a transition from the braided pattern to the meandering one. The analysis of the simulation process indicated that (1) a decrease in the flow discharge and sediment supply can lead to the transition; and (2) the riparian vegetation helps stabilize the cut bank and bar surface but it is not key to the transition. The results are in agreement with the criterion proposed in previous research, confirming the 2D numerical model's potential in predicting the transition between different channel patterns and improving our understanding of the fluvial process. It was concluded that, further studies are needed to research the fundamental equation that governs the evolution of alluvial river, which has not been fully understood to ensure the availability of the numerical model.

The research presented in [6] aimed to present improvements in the computational efficiency of a 2D hydrodynamic model that can lead to a significant bearing on its engineering applications. In the presented study, an improved 2D shallow water dynamic model applying a local time stepping (LTS) algorithm was established using a finite volume method (FVM) on a triangular mesh. Results from the simulation and analysis of three canonical test cases showed that the LTS scheme has a satisfactory ability for adapting real complex applications and long simulations while meeting the required accuracy. The following conclusions were drawn: (1) based on the FVM for unstructured grids, the implemented LTS algorithm improved the computational efficiency of the model, while satisfying water conservation conditions. In the anti-symmetric dam break case, a speedup ratio of 2.1 was achieved, which saved 53% in execution time. The speedup ratio of the non-flat bottom dam break case was 1.3, which represented a shortening of 26% in the calculation time. The numerical simulation of the navigable flow of the river reach between the Three Gorges and Gezhouba Dams achieved a speedup ratio of 1.9, which represented a saving of 49% in modeling time; (2) the proportions of coarse to refined meshes on the acceleration effect of the LTS algorithm were noticeable. It was evident that a higher speedup ratio was obtained when the proportion of the refined mesh was minimized. When the proportion of the refined mesh was high, the acceleration effect was not significant. It is clear that the LTS algorithm is best suited to situations in which refinement is only required in small regions; and (3) when using the LTS algorithm on non-uniform unstructured grids, the larger the grid scale difference, the more obvious the grid layering became. This led to increased acceleration effects. However, computational accuracy was slightly impaired by excessive differences in grid mesh size. These results can provide technical guidelines for reducing computational time for 2D hydrodynamic models on non-uniform unstructured grids.

In [7], the main goal was to use various numerical investigations to figure out the response of coastal tides on the west coast of Korea (WCK) to the open-boundary conditions and bottom roughness using an open source computational fluids dynamics tool—the TELEMAC model. Three well-known assimilated tidal models were used to obtain a detailed tidal forcing at open boundaries—the finite element solution (FES2014); the Oregon State University TOPEX/Poseidon Global Inverse Solution Tidal Model (TPXO9.1); and the National Astronomical Observatory of Japan (NAO.99Jb). It was shown that there were no significant differences between the responses in tidal amplitudes in the WCK induced by three open-boundary conditions obtained from three assimilated tidal models. In addition, the numerical simulation of the tidal flow in the WCK should not use a uniform bottom roughness coefficient. Due to the complicated bathymetry, indented coastlines and bed variability of the WCK, it caused strong local effects on the tides in this region. Therefore, a non-uniform bottom roughness should be applied to the modeling whereby the smallest value can be applied for Incheon, a larger value for Mokpo, and the largest value for Gunsan. The largest value of the bottom roughness coefficient was applied to the Gunsan region because its bed form was more variable than in other regions. The numerical results showed that the accuracy of the modeling of the tidal elevation around the WCK was strongly dependent on the bottom roughness rather than the offshore tidal boundary conditions. Moreover, the numerical results can provide not only a better fit to the observations but also higher spatial resolutions in comparison to the results obtained from assimilated tidal models around the WCK. However, it should be noted that the numerical results obtained from this study were still limited due to the coarse resolution (30 arcs/s) of the bathymetry obtained from GEBCO2014—which was not sufficient to capture the real geometries whose sizes were less than such resolution. Therefore, a further study with a higher resolution is necessary in order to obtain a more precise prediction of the tidal current and its elevation around the WCK. Furthermore, the wind forcing on the sea surface and the tidal energy dissipation should be taken into account.

In [8], the one-dimensional Boussinesq equations were numerically solved using the MacCormack as well as the Dissipative two–four finite difference schemes for the simulation of hydraulic jump formation in a horizontal rectangular open channel and for upstream Froude numbers Fr in the range of 2.44–5.38. The governing equations were enriched with additional terms if compared to the Saint-Venant equations, to account for the non-hydrostatic pressure distribution in the regime of rapidly varied flow. Terms related to the energy loss and the gravity forces were also included. The initial condition was a steady supercritical gradually varied flow along the whole length of the modeled channel. The upstream and downstream boundary conditions regarding the flow depth remained constant during the iteration process and equal to the values measured in experiments. The method of specified intervals was used for the calculation of the velocity at the downstream end, assuming that the positive characteristic through a point does not intersect with already established grid points. The variable time step was used in every iteration according to the CFL (Courant–Friedrichs–Lewy) stability criterion, along with artificial viscosity for the smoothing of the oscillations occurring in the jump. The computational results compared well with the experiments since the specific force was computed from the depth and mean velocity at both ends of the hydraulic jump with acceptable tolerance, and the mass conservation equation was verified for all numerical schemes and all test cases. From such a model, one can determine the sequent depth ratio as well as the length of the jump, results that are useful in the design of stilling basins (geometrical properties). Given a stilling basin with a known inflow Froude number and flow depth, the engineer must decide the end sill dimensions and the basin length, so that the hydraulic jump is contained in the stilling basin. Finally, from the comparison of the numerical results and experiments, it can be concluded that the aforementioned numerical modeling schemes can predict the basic features of the classical hydraulic jump with acceptable accuracy.

In [9], the authors introduced several innovations, including the use of the non-linear Krylov accelerator in open-channel flows, an evolutionary domain algorithm and the use of CasADi to solve steady 1D flows using the Saint-Venant model equations. These improvements led to an algorithm that is able to quickly solve steady open-channel flows. Therefore, optimization problems and uncertainty analyses that require many evaluations become more tractable. An original algorithm was implemented in order to significantly improve the computation time of a steady 1D open-channel flow problem. This includes two main optimizing strategies: a non-linear Krylov accelerator and an evolutionary domain algorithm. This new algorithm was validated against the academic benchmarks of flows over a bump. The results showed a good agreement between the numerical and analytical values. The performance of the suggested algorithm was evaluated against the non-linear optimization software CasADi. It showed good scalability properties. Indeed, the execution time of the proposed algorithm linearly evolves with the number of nodes.

This is not the case with other techniques when the mesh is refined and/or when the number of nodes is increased. Finally, the capabilities of the proposed algorithm were validated on a real-world case. The optimized algorithm was used in order to quickly compute the initial condition required by the operational model for the Romanche River in France. The technique was able to provide a steady-state solution to the unsteady model in a very short period of time.

In [10], the performances of several modeling approaches were compared in order to evaluate their results and computational requirements in a transient river flow event in a reach of the Ebro river (Spain) that includes a reservoir covering a large area. A 2D distributed shallow water model solved over a triangular grid and a 1D shallow water model was used to discretize the full domain. Additionally, an aggregated volume balance model was implemented to model a reservoir region in order to allow computational saving. This led to a coupled 1D–0D model. Finally, a proportional–integral–derivative (PID) control algorithm was implemented as a regulation technique at the dam location and combined with both the 1D model and the 1D–0D model. From the comparison of the performances of the 2D and 1D models, it was concluded that the results of the 1D model for the recent flooding events at the considered Ebro River reach were very similar to those provided by the 2D model. The water level and discharge data predicted by both models follow the same trend. The cross-sections used to build the 1D model computational mesh were carefully located to reproduce the river curvature in detail, which is important to obtain a realistic evolution of the hydraulic variables. This effort is justified by the immense computational saving that the use of the 1D model offers, as long as there is no interest in representing the floodplain flow that the 1D model does not take into account. The coupling of the 1D model for the river flow at the upstream reach and the 0D model for the reservoir (1D–0D model) offers results very similar to those from the full 1D model. There was some lag due to the instantaneous propagation of the hydrograph in the reservoir assumed by the 0D model but this is acceptable considering the computational savings that the use of this model implies compared to the full 1D model. The computational times observed with the 1D–0D model justifies the use of this combined approach. Therefore, the coupling of a 0D model for the reservoir with the 2D model for the upstream river reach was envisaged as future work since this will lead to high computational savings, something very positive for simulations with 2D models as well as the possibility to simulate the floodplain flow behavior. The PID control algorithm was implemented with the objective to ensure a fixed-surface water level at the dam. The results showed that this target level value was never reached despite the time variable discharge, which means that the implementation of the control algorithm is a correct security measure to avoid exceeding certain levels in the reservoir. It will be convenient in the future to implement an algorithm that takes into account more realistic and complex objectives.

Author Contributions: A.I.D. and I.K.N. conceived, designed, and wrote the editorial. All authors have read and agreed to the published version of the manuscript.

Funding: This research received no external funding.

Acknowledgments: Thanks to all of the contributions to the Special Issue, the time invested by each author, as well as to the anonymous reviewers and editorial managers who have contributed to the development of the articles in this Special Issue.

Conflicts of Interest: The authors declare no conflict of interest.

References

1. Brocchini, M.; Dodd, N.Nonlinear Shallow Water Equation Modeling for Coastal Engineering. *J. Waterw. Port Coast. Ocean. Eng.* **2008**, *134*, 104. [CrossRef]
2. Delis, A.I.; Kampanis, N.A. Numerical flood simulation by depth averaged free surface flow models. In *Environmental Systems, Encyclopedia of Life Support Systems*; Sydov, A., Ed.; Developed under the Auspices of the UNESCO; Eolss Publishers: Oxford, UK, 2009.
3. Xing, Y. Numerical Methods for the Nonlinear Shallow Water Equations. In *Handbook of Numerical Analysis*; Elsevier: Amsterdam, The Netherlands, 2017; Volume 18, pp. 361–384.
4. Ginting, B.M.; Mundani, R.-P. Comparison of Shallow Water Solvers: Applications for Dam-Break and Tsunami Cases with Reordering Strategy for Efficient Vectorization on Modern Hardware. *Water* **2019**, *11*, 639. [CrossRef]
5. Yang, S.; Xiao, Y. 2D Numerical Modeling on the Transformation Mechanism of the Braided Channel. *Water* **2019**, *11*, 2030. [CrossRef]
6. Yang, X.; An, W.; Li, W.; Zhang, S. Implementation of a Local Time Stepping Algorithm and Its Acceleration Effect on Two-Dimensional Hydrodynamic Models. *Water* **2020**, *12*, 1148. [CrossRef]
7. Nguyen, V.T.; Lee, M. Effect of Open Boundary Conditions and Bottom Roughness on Tidal Modeling around the West Coast of Korea. *Water* **2020**, *12*, 1706. [CrossRef]
8. Retsinis, E.; Papanicolaou, P. Numerical and Experimental Study of Classical Hydraulic Jump. *Water* **2020**, *12*, 1766. [CrossRef]
9. Goffin, L.; Dewals, B.; Erpicum, S.; Pirotton, M.; Archambeau, P. An Optimized and Scalable Algorithm for the Fast Convergence of Steady 1-D Open-Channel Flows. *Water* **2020**, *12*, 3218. [CrossRef]
10. Echeverribar, I.; Vallés, P.; Mairal, J.; García-Navarro, P. Efficient Reservoir Modelling for Flood Regulation in the Ebro River (Spain). *Water* **2021**, *13*, 3160. [CrossRef]

Article

Comparison of Shallow Water Solvers: Applications for Dam-Break and Tsunami Cases with Reordering Strategy for Efficient Vectorization on Modern Hardware

Bobby Minola Ginting * and Ralf-Peter Mundani

Chair for Computation in Engineering, Technical University of Munich, Arcisstr. 21, D-80333 Munich, Germany; mundani@tum.de

* Correspondence: bobbyminola.ginting@tum.de; Tel.: +49-89-289-23044

Received: 13 February 2019; Accepted: 21 March 2019; Published: 27 March 2019

Abstract: We investigate in this paper the behaviors of the Riemann solvers (Roe and Harten-Lax-van Leer-Contact (HLLC) schemes) and the Riemann-solver-free method (central-upwind scheme) regarding their accuracy and efficiency for solving the 2D shallow water equations. Our model was devised to be spatially second-order accurate with the Monotonic Upwind Scheme for Conservation Laws (MUSCL) reconstruction for a cell-centered finite volume scheme—and be temporally fourth-order accurate using the Runge–Kutta fourth-order method. Four benchmark cases of dam-break and tsunami events dealing with highly-discontinuous flows and wet–dry problems were simulated. To this end, we applied a reordering strategy for the data structures in our code supporting efficient vectorization and memory access alignment for boosting the performance. Two main features are pointed out here. Firstly, the reordering strategy employed has enabled highly-efficient vectorization for the three solvers investigated on three modern hardware (AVX, AVX2, and AVX-512), where speed-ups of 4.5–6.5× were obtained on the AVX/AVX2 machines for eight data per vector while on the AVX-512 machine we achieved a speed-up of up to 16.7× for 16 data per vector, all with singe-core computation; with parallel simulations, speed-ups of up to 75.7–121.8× and 928.9× were obtained on AVX/AVX2 and AVX-512 machines, respectively. Secondly, we observed that the central-upwind scheme was able to outperform the HLLC and Roe schemes 1.4× and 1.25×, respectively, by exhibiting similar accuracies. This study would be useful for modelers who are interested in developing shallow water codes.

Keywords: central-upwind; efficiency; finite volume; HLLC; modern hardware; Roe; shallow water equations; vectorization

---

1. Introduction

Dam-break or tsunami flows cause not only potential dangers to human life, but also great losses of property. These phenomena can be triggered by some natural hazards, such as earthquakes or heavy rainfall. When a dam breaks, a large amount of water is released instantaneously from the dam and will propagate rapidly to the downstream area. Similarly, tsunami waves flowing rapidly from the ocean bring a large volume of water to coastal areas, which endangers human life as well as damages infrastructure. Since natural hazards have very complex characteristics, in terms of the spatial and temporal scales, they are quite difficult to predict precisely. Therefore, it is highly important to study the evolution of such flows as a part of a disaster management, which will be useful for the related stakeholders in decision-making. Such study can be done by developing a mathematical model based on the 2D shallow water equations (SWEs).

Recent numerical models of the 2D SWEs rely, almost entirely, on the computations of (approximate) Riemann solvers, particularly in the applications of the high-resolution Godunov-type methods. The simplicity, robustness, and built-in conservation properties of the Riemann solvers, such as the Roe and HLLC schemes, had led to many successful applications in shallow flow simulations, see [1–5], among others. Highly discontinuous flows, including transcritical flows, shock waves and moving wet–dry fronts were accurately simulated.

Generally speaking, a scheme can be regarded as a class of Riemann solvers if it is proposed based on a Riemann problem. The Roe scheme was originally devised by [6] and was proposed to estimate the interface convective fluxes between two adjacent cells on a spatially-and-temporally discretized computational domain by linearizing the Jacobian matrix of the partial differential equations (PDEs) with regard to its left and right eigenvectors. This linearized part contributes to the computation of the convective fluxes of the PDEs, as a flux difference for the average value of the considered edge taken from its two corresponding cells. Since the eigenstructure of the PDEs—which leads to an approximation of the interface value in connection with the local Riemann problem—must be known in the calculation of the flux difference, the Roe scheme is regarded as an approximate Riemann solver.

More than 20 years later, Toro [7] then developed a new approximate Riemann solver—HLLC scheme—to simulate shallow water flows, which was an extended version of the Harten-Lax-van Leer (HLL) scheme proposed in [8]. In the HLL scheme, the solution is approximated directly for the interface fluxes by dividing the region into three parts: left, middle, and right. Both the left and right regions correspond to the values of the two adjacent cells, whereas the middle region consists of a single value separated by intermediate waves. One major flaw of the HLL scheme is related to both contact discontinuities and shear waves leading to a missing contact (middle) wave. Therefore, Toro [7] fixed this scheme in the HLLC scheme by including the computation of the middle wave speed that now the solution is divided into four regions. There are several ways to calculate the middle wave speed, see [9–11]. All the calculations deal with the eigenstructure of the PDEs, which is related to the local Riemann problem, and obviously, this brings the HLLC scheme back to a class of Riemann solvers.

Opposite to the Riemann solvers, Kurganov et al. [12] proposed the central-upwind (CU) method as a Riemann-solver-free scheme, in which the eigenstructure of the PDEs is not required to calculate the convective fluxes. Instead, the local one-sided speeds of propagation at every edge, which can be computed in a straight-forward manner, are used. This scheme has been proven to be sufficiently robust and at the same time can satisfy both the well-balanced and positivity preserving properties, see [13–15].

To solve the time-dependent SWEs, all the aforementioned schemes must be temporally discretized either by using an implicit or an explicit time stepping method. Despite its simplicity, the latter may, however, suffer from a stability computational issue particularly when simulating a very low water on a very rough bed [16,17]. The former is unconditionally stable and even is very flexible to use a large time step. However, the computation is admittedly complex. Another way that can be used to overcome the stability issue of the explicit method and to avoid the complexity of the implicit method—is to perform a high-order explicit method, such as the Runge–Kutta high-order scheme. This method is more stable than the explicit method, while the computation remains simple and acceptably cheap as that of the explicit method.

As the high-order time stepping method is now considered, the selection of solvers included in models must be taken into careful consideration, since such solvers—which are the most expensive part in SWEs simulations—need to be computed several times in a single time step. For example, the Runge–Kutta fourth-order (RKFO) method requires the updating of a solver four times to determine the value at the subsequent time step. The more complex the algorithm of a solver is, the more CPU time one obtains.

Nowadays, performing SWE simulations is becoming more and more common on modern hardware/CPUs towards high-performance computing (HPC) using advanced features such as

AVX, AVX2, and AVX-512, which support the algorithm vectorization for executing some operations in a single instruction—known as single instruction multiple data (SIMD)—so that a significant computation speed-up can be achieved. Vectorization on such modern hardware employs vector instructions, which can dramatically outperform scalar instructions, thus being quite important for having more efficient computations. Among the other compilers' optimizations, vectorization can even be regarded as the common ways for utilizing vector-level parallelism, see [18,19]. Such a speed-up, however, can only exist if the algorithm formulation is suitable for vectorization instructions either automatically (by compilers) or manually (by users) [20].

Typically, there are three classes of vectorization: auto vectorization, guided vectorization, and low-level vectorization. The first type is the easiest one utilizing the ability of the compiler to automatically detect loops, which have a potential to be vectorized. This can be done at compiling time, e.g., using the optimization flag `-O2` or higher. However, some typical problems, e.g., non-contiguous memory access and data-dependency, make vectorization difficult. For this, the second type may be a solution utilizing some compiler hints/pragmas and array notations. This type may successfully vectorize the loops that cannot be auto-vectorized by the compiler. However, if not used carefully, it gives no significant performance or even the results can be wrong. The last type is probably the hardest one since it requires deep-knowledge about intrinsics/assembly programming and vector classes, thus not so popular.

Especially in simulating complex phenomena such as dam-break or tsunami flows as part of disaster planning, accurate results are obviously of particular interest for modelers. However, focusing only on numerical accuracy but ignoring performance efficiency is no longer an option. For instance, in addition to relatively large-sized domains, most of real dam-break and tsunami simulations require performing long real-time computations, e.g., days or even up to weeks. Wasting the performance either due to the complexity level of the solver selected or the code's inability to utilize the vectorization, is thus undesirable. This becomes our focus in this paper. We compare three common shallow water solvers (HLLC, Roe, and CU schemes) here, where two main findings are pointed out. Firstly, to enable highly-efficient vectorization for all solvers on all the aforementioned hardware, we employ a reordering strategy that we have recently applied in [21]. This strategy supports guided vectorization and memory access alignment for the array loops attempted in the SWEs' computations, thus boosting the performance. Secondly, we observe that the CU scheme is capable of outperforming the performance of the HLLC and Roe schemes by exhibiting similar accuracies. These findings would be useful for modelers as a reference to select the numerical solvers to be included in their models as well as to optimize their codes for vectorization.

Some previous studies reporting about vectorization of shallow water solvers are noted here. In [20], the augmented Riemann solver implemented in a source code Geo Conservation Laws (GeoCLAW) was vectorized using a low-level vectorization with SSE4 and AVX intrinsics. The average speed-up factors of 2.7× and 4.1× (both with single-precision arithmetic) were achieved with SSE4 and AVX machines, respectively. Also using GeoCLAW, the augmented Riemann solver was vectorized in [22] by changing the data layouts from arrays of structs (AoS) to structs of arrays (SoA), thus requiring a considerably huge task for rewriting the code—and then applying a guided vectorization with `!$omp simd`. The average speed-up factors of 1.7× and 4.4× (both with double-precision arithmetic) were achieved with AVX2 and AVX-512 machines, respectively. In [23], the split HLL Riemann solver was vectorized and parallelized for the flux computation and state computation modules of the SWEs employing low-level vectorization with SSE4 and AVX intrinsics. To the best of our knowledge, this is the first attempt to report the efficiency comparisons of common solvers (both Riemann and non-Riemann solvers) regarding the vectorization on the three modern hardwares without having to perform complex intrinsic functions or to require a lot of work for rewriting the code. We use here an in-house code of the first-author—numerical simulation of free surface shallow water 2D (NUFSAW2D). Some successful applications were shown using NUFSAW2D for varying shallow water-type simulations, e.g., dam-break cases, overland flows, and turbulent flows, see [17,21,24,25].

This paper is organized as follows. The governing equations and the numerical model are briefly explained in Section 2. An overview of data structures in our code is presented in Section 3. The model verifications against the benchmark cases and its performance evaluations are given in Section 4. Finally, conclusions are given in Section 5.

2. Governing Equations and Numerical Models

The 2D SWEs are written in conservative form according to [26] as

$$\frac{\partial \mathbf{W}}{\partial t} + \frac{\partial \mathbf{F}}{\partial x} + \frac{\partial \mathbf{G}}{\partial y} = \mathbf{S}_b + \mathbf{S}_f \,, \tag{1}$$

where the vectors $\mathbf{W}$, $\mathbf{F}$, $\mathbf{G}$, $\mathbf{S}_b$, and $\mathbf{S}_f$ are expressed as

$$\mathbf{W} = \begin{bmatrix} h \\ hu \\ hv \end{bmatrix} , \quad \mathbf{F} = \begin{bmatrix} hu \\ huu + \dfrac{gh^2}{2} \\ hvu \end{bmatrix} , \quad \mathbf{G} = \begin{bmatrix} hv \\ huv \\ hvv + \dfrac{gh^2}{2} \end{bmatrix} ,$$
$$\mathbf{S}_b = \begin{bmatrix} 0 \\ -gh\dfrac{\partial z_b}{\partial x} \\ -gh\dfrac{\partial z_b}{\partial y} \end{bmatrix} , \quad \mathbf{S}_f = \begin{bmatrix} 0 \\ -g\, h\dfrac{n_m^2\, u\, \sqrt{u^2+v^2}}{h^{4/3}} \\ -g\, h\dfrac{n_m^2\, v\, \sqrt{u^2+v^2}}{h^{4/3}} \end{bmatrix} . \tag{2}$$

The water depth, velocities in x and y directions, gravity acceleration, bottom elevation, and Manning coefficient are denoted by h, u, v, g, z_b, and n_m, respectively. Using a cell-centered finite volume (CCFV) method, Equation (1) is spatially discretized over a domain Ω as

$$\frac{\partial}{\partial t}\iint_{\Omega} \mathbf{W} d\Omega + \iint_{\Omega} \left(\frac{\partial \mathbf{F}}{\partial x} + \frac{\partial \mathbf{G}}{\partial y} \right) d\Omega = \iint_{\Omega} (\mathbf{S}_b + \mathbf{S}_f) d\Omega \,. \tag{3}$$

Applying the Gauss divergence theorem, the convective fluxes of Equation (3) can be transformed into a line-boundary integral Γ as

$$\frac{\partial}{\partial t}\iint_{\Omega} \mathbf{W} d\Omega + \oint_{\Gamma} (\mathbf{F}\, n_x + \mathbf{G}\, n_y)\, d\Gamma = \iint_{\Omega} (\mathbf{S}_b + \mathbf{S}_f) d\Omega \,, \tag{4}$$

leading to a flux summation for the convective fluxes by

$$\oint_{\Gamma} (\mathbf{F} n_x + \mathbf{G} n_y)\, d\Gamma \approx \sum_{i=1}^{N} (\mathbf{F}\, n_x + \mathbf{G}\, n_y)_i \, \Delta L_i \,, \tag{5}$$

where n_x and n_y are the normal vectors outward Γ, N is the total number of edges for a cell, and ΔL is the edge length. We will investigate the accuracy and efficiency of the three solvers for solving Equation (5). The in-house code NUFSAW2D used here implements the modern shock-capturing Godunov-type model, which supports the structured as well as unstructured meshes by storing the average values in each cell-center. Here we use structured rectangular meshes, hence N = 4. The second-order spatial accuracy was achieved with the MUSCL method utilizing the MinMod limiter function to enforce the monotonicity in multiple dimensions. The bed-slope terms were computed using a Riemann-solution-free technique, with which the bed-slope fluxes can be computed separately from the convective fluxes, thus giving a fair comparison for the three aforementioned solvers. The friction terms were treated semi-implicitly to ensure stability for wet–dry simulations. The RKFO method is now applied to Equation (4) as

$$\begin{aligned} \mathbf{W}^{p=0} &= \mathbf{W}^{t}\,, \qquad \texttt{for}\ \ p = 1, \ldots, 4\ \ \texttt{then} \\ \mathbf{W}^{p} &= \mathbf{W}^{p=0} + \alpha_p \left[-\frac{\Delta t}{A} \sum_{i=1}^{4} \left(\mathbf{F}\, n_x + \mathbf{G}\, n_y\right)_i \Delta L_i + \Delta t \iint_{\Omega} (\mathbf{S}_b + \mathbf{S}_f)\, d\Omega \right]^{p-1}, \\ \mathbf{W}^{t+1} &= \mathbf{W}^{p=4}\,, \end{aligned} \tag{6}$$

where A is the cell area, Δt is the time step, α_p is the coefficient being $1/4$, $1/3$, $1/2$, and 1 for p = 1–4, respectively. The numerical procedures for Equations (4) and (6) are given in detail in [17,25,26], thus are not presented here.

3. Overview of Data Structures

3.1. General

Here we explain in detail how the data structures of our code are designed to advance the solutions of Equation (6). Note this is a typical data structure used in many shallow water codes (with implementations of modern finite volume schemes). As shown in Figure 1, a domain is discretized into several sub-domains (rectangular cells). We call this step the pre-processing stage. Each cell now consists of the values of z_b and n_m located at its center. Initially, the values of h, u, and v are given by users at each cell-center.

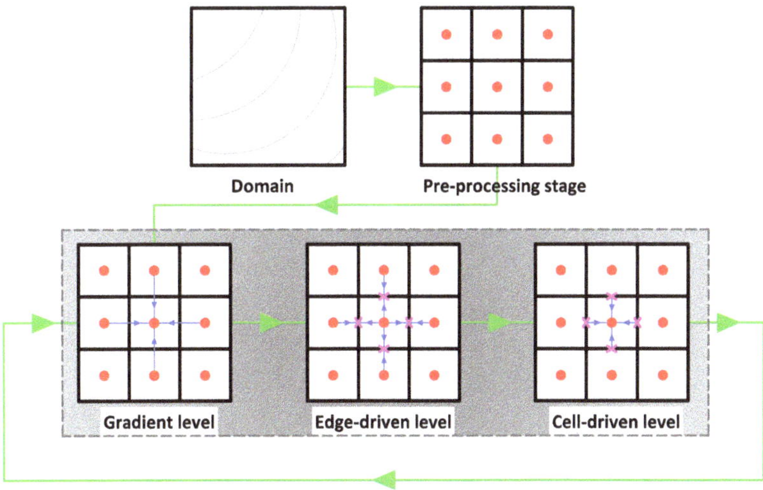

Figure 1. Typical process in shallow flow modeling (with implementations of modern finite volume schemes).

As our model employs a reconstruction process to spatially achieve second-order accuracy with the MUSCL method, it requires the gradient values at cell-center. Therefore, these gradient values must firstly be computed. This step is called the gradient level. Hereafter, one requires to calculate the values at each edge using the values of its two corresponding cell-centers. This stage is then called the edge-driven level. In this level, a solver, e.g., HLLC, Roe, or CU scheme, is required to compute the non-linear values of **F** and **G** at edges. Prior to performing such a solver, the aforementioned reconstruction process with the MUSCL method was employed. Note the values of $\mathbf{S}_b$ are also computed at the edge-driven level. After the values of all edges are known, the solution can be advanced for the subsequent time level by also computing the values of $\mathbf{S}_f$. For example, the solutions of **W** at the subsequent time level for a cell-center are updated using the **F**, **G**, and $\mathbf{S}_b$ values from its four corresponding edges—and using $\mathbf{S}_f$ values located at the cell-center itself. We call this stage the cell-driven level.

Note that the edge-driven level is the most expensive stage among the others; one should thus pay extra attention to its computation. We also point out here that we apply the computation for the edge-driven level in an edge-based manner rather than in a cell-based one, namely we compute the edge values only once per single calculation level. Therefore, one does not need to save the values of $\left[\sum_{i=1}^{N}\left(\mathbf{F}\,n_x+\mathbf{G}\,n_y\right)_i\Delta L_i\right]$ in arrays for each cell-center; only the values of $\left[\left(\mathbf{F}\,n_x+\mathbf{G}\,n_y\right)_i\Delta L_i\right]$ are saved corresponding to the total number of edges, instead. The values of an edge are only valid for one adjacent cell—and such values are simply multiplied by (-1) for another cell. It is now a challenging task to design an array structure that can ease vectorization and exploit memory access alignment in both the edge-driven and cell-driven levels.

3.2. Cell-Edge Reordering Strategy for Supporting Vectorization and Memory Access Alignment

We focus our reordering strategy here on tackling the two common problems for vectorization: non-contiguous memory access and data-dependency. Regarding the former, a contiguous array structure is required to provide contiguous memory access giving an efficient vectorization. Typically, one finds this problem when dealing with an indirect array indexing, e.g., using `x(y(i))` forces the compiler to decode `y(i)` for finding the memory reference of `x`. This is also a typical problem for a non-unit strided access to array, e.g., incrementing a loop by a scalar factor, where non-consecutive memory locations must be accessed in the loop. The vectorization is sometimes still possible for this problem type. However, the performance gain is often not significant. The second problem relates to usage of arrays identical to the previous iteration of the loop, which often destroy any possibility for vectorization, otherwise a special directive should be used.

See Figure 2, for advancing the solution of **W** in Equation (1) for `k`, one requires **F**, **G**, and $\mathbf{S}_b$ from `i`, where `i` = `index_function(j)` and [`j` ← `1-4`]—and $\mathbf{S}_f$ from `k` itself. Opting `index_function` as an operator for defining `i` leads to a use of an indirect reference in a loop. This is not desired since it may avoid the vectorization. This may be anticipated by directly declaring `i` into the same array to that of `k`, e.g., `W(k)` ← [`W(k+m)`,`W(k-m)`,`W(k+n)`,`W(k-n)`], where `m` and `n` are scalar. This, however, leads to a data-dependency problem that makes vectorization difficult.

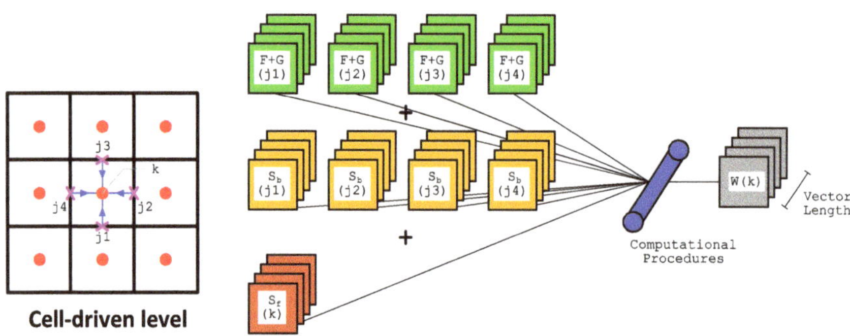

Figure 2. Vectorization for advancing the solution in the cell-driven level.

To avoid these problems, we have designed a cell-edge reordering strategy, see Figure 3, where the loops with similar computational procedures are collected to be vectorized. Note that this strategy is only applied once at the pre-processing stage in Figure 1. The core idea of this strategy is to build contiguous array patterns between edges and cells for the edge-driven level as well as between cells and edges for the cell-driven level. We point out here that we only employ 1D array configuration in NUFSAW2D, so that the memory access patterns are straightforward, thus easing unit stride and conserving cache entries. The first step is to devise the cell numbering following the Z-pattern, which is intended for the cell-driven level. Secondly, we design the edge numbering for the edge-driven level by classifying the edges into two types: internal and boundary edges in the most contiguous way; the former is the edges that have two neighboring cells (e.g., edges 1–31), whereas the latter is the edges with only one corresponding cell (e.g., edges 32–49). The reason for this classification is the computational complexity between the internal and boundary edges differs from each other, e.g., (1) no reconstruction process is required for the latter, thus having less CPU time than the former—and (2) due to corresponding to two neighboring cells, the former accesses more memories than does the latter; declaring all edges only in one single loop-group therefore deteriorates the memory access patterns, thus decreasing the performance.

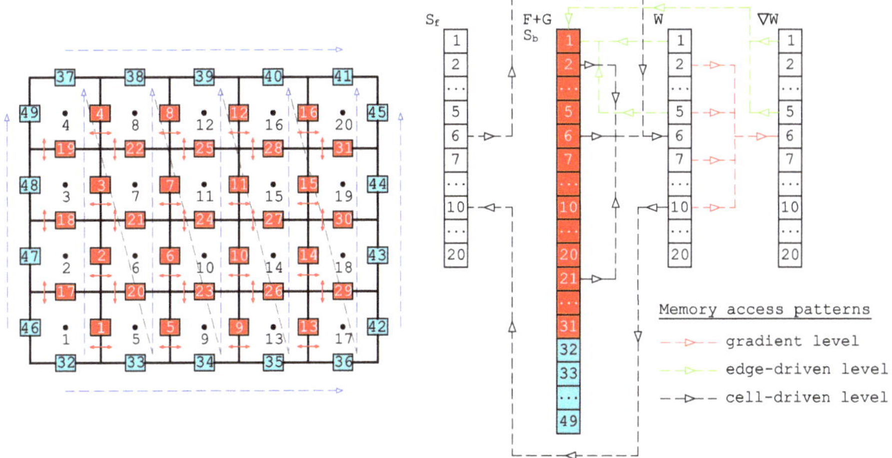

Figure 3. Cell-edge reordering strategy [21] and an example of memory access patterns.

For the sake of clarity, we write in Algorithm 1 the pseudo-code of the model's `SUBROUTINE` employed in NUFSAW2D. Note that Algorithm 1 is a typical form applied in many common and popular shallow water codes. First, we mention that `seg_x = 5`, `seg_y = 4`, and `Ncells = 20` according to Figure 3, where `seg_x`, `seg_y`, and `Ncells` are the total number of domain segments in x and y directions, and the total number of cells, respectively. We now explain the `SUBROUTINE gradient`. The cells are now classified into two groups: internal and boundary cells. Internal cells, e.g., cells 6, 7, 10, 11, 14, and 15 are cells whose gradient computations require accessing two cell values in each direction. For example, computing the x-gradient of **W** of cell 6 needs the values of **W** of cells 2 and 10; this is denoted by $[\nabla \mathtt{W_x(6)} \leftarrow \mathtt{W(2),W(10)}]$ and similarly $[\nabla \mathtt{W_y(6)} \leftarrow \mathtt{W(5),W(7)}]$. Boundary cells, e.g., cells 1–4, 5, 8, 9, 12, 13, 16, and 17–20, are cells affiliated with boundary edges. These cells may not always require accessing two cell values in each direction for the gradient computation, e.g., $[\nabla \mathtt{W_x(8)} \leftarrow \mathtt{W(4),W(12)}]$ but $[\nabla \mathtt{W_y(8)} \leftarrow \mathtt{W(7),W(8)}]$ showing that a symmetric boundary condition is applied to cell 8 in y direction. Considering the fact that the total number of internal cells is significantly larger than that of boundary cells, we group the internal cells into a single loop and distinguish them from the boundary cells, see Algorithm 2.

Algorithm 1 Typical algorithm for shallow water code (within the Runge–Kutta fourth-order (RKFO) method's framework)

```
for t = 1 ← [total number of time step] do
  ! within the RKFO method from [p=1] to [p=4]
  for p = 1 ← 4 do
    CALL gradient
      → compute gradient
    CALL edge-driven_level
      → compute MUSCL_method
      → compute bed_slope
      → compute shallow_water_solver
    CALL cell-driven_level
      → compute friction_term
      → compute update_variables
  end for
end for
```

Algorithm 2 Pseudo-code for `SUBROUTINE gradient`

```
1:  for k=1 ← [seg_x-2] do
2:    l = (seg_y+2)+(k-1)*seg_y
3:    !$omp simd simdlen(VL) aligned(∇W_x,∇W_y :Nbyte)
4:    for i=1 ← [1+seg_y-3] do
5:      .........
6:      ∇W_x(i) ← W(i-seg_y),W(i+seg_y)  ;  ∇W_y(i) ← W(i-1),W(i+1)
7:    end for
8:  end for
9:  !$omp simd simdlen(VL) aligned(∇W_x,∇W_y :Nbyte)
10: for i=1 ← [seg_y] do
11:   j=Ncells-seg_y+i
12:   i1=i-1 OR i1=i     ;     i2=i+1 OR i2=i
13:   i3=j-1 OR i3=j     ;     i4=j+1 OR i4=j
14:   .........
15:   ∇W_x(i) ← W(i),W(i+seg_y)  ;  ∇W_y(i) ← W(i1),W(i2)
16:   ∇W_x(j) ← W(j-seg_y),W(j)  ;  ∇W_y(j) ← W(i3),W(i4)
17: end for
18: !=== This loop is not vectorized due to non-unit strided access ===!
19: for i=1 ← [seg_x-2] do
20:   j=i*seg_y+1        ;     k=(i+1)*seg_y
21:   i1=j-1 OR i1=j     ;     i2=j+1 OR i2=j
22:   i3=k-1 OR i3=k     ;     i4=k+1 OR i4=k
23:   .........
24:   ∇W_x(j) ← W(j-seg_y),W(j+seg_y)  ;  ∇W_y(j) ← W(i1),W(i2)
25:   ∇W_x(k) ← W(k-seg_y),W(k+seg_y)  ;  ∇W_y(k) ← W(i3),W(i4)
26: end for
```

Algorithm 2 shows three typical loops in the `SUBROUTINE gradient`. The first loop (lines 1–8) is designed sequentially with a factor of `seg_x-2` for its outer part to exclude all boundary cells. For its inner part, this loop is constructed based on the outer loop in a contiguous way, thus making vectorization efficient. Each element of array ∇W_x accesses two elements from array `W` with the farthest alignment of `seg_y`, while each element of array ∇W_y also accesses two elements of array `W` but only with the farthest alignment of `1`. The second loop (lines 10–17) is also designed similarly to the first one, but since this loop includes boundary cells, each element of arrays ∇W_x and ∇W_y only accesses one array with the farthest alignment of `seg_y` and `1`, respectively—whereas the other elements from array `W` required are contiguously accessed by each element of both ∇W_x and ∇W_y. Note in our implementation, none of these two loops can be auto-vectorized by the compiler. Therefore, we apply a guided vectorization with OpenMP directive instead of the Intel one, namely `!$omp simd simdlen(VL) aligned(var1,var2,... :Nbyte)`; this will be explained later in Section 4.5. The third loop (lines 19–26) is designed for the rest cells, which are not included in the previous two loops. This loop is not devised in a contiguous manner, thus disabling auto vectorization or, although a guided vectorization is possible, it still does not give any significant performance improvement due to non-unit strided access. Despite being unable to be vectorized, the third loop does not significantly decrease the performance of our model for the entire simulation as it only has an array dimension of `2*[seg_x-2]` (quite small compared to the other two loops).

We now discuss the `SUBROUTINE edge-driven_level` and sketch it in Algorithm 3. Note for the sake of brevity, only the pseudo-code for internal edges is represented in Algorithm 3; for boundary edges, the pseudo-code is similar but computed without `MUSCL_method`. The first loop corresponds to the edges 1–16 and the second one to the edges 17–31. In the first loop (lines 1–7), each flux computation accesses the array with the farthest alignment of `seg_y`, whereas the arrays are designed in the second loop (lines 8–17) to have contiguous patterns. Every edge has a certain pattern for its two corresponding cells, where no data-dependency exists, thus enabling an efficient vectorization. Note with this pattern, both loops can be auto-vectorized; however, we still implement a guided vectorization as it gives a better performance.

Finally, we sketch the `SUBROUTINE cell-driven_level` in Algorithm 4. Again, for the sake of brevity only the pseudo-code for internal cells is given. Similar to the internal cell in the `SUBROUTINE gradient`, the loop is designed sequentially with a factor of `seg_x-2` for the outer part. In the inner part the arrays access patterns are, however, different to those of the gradient computation, where `W` accesses `F`, `G`, and S_b from the corresponding edges—and S_f from the corresponding cell; in other words, more array accesses are required in this loop. Nevertheless, the vectorization gives a significant performance improvement since the array accesses patterns are contiguous. However, there is a part that cannot be vectorized in this cell-driven level due to non-unit strided access, similar to that shown in Algorithm 2. Again, since the dimension of this non-vectorizable loop is considerably smaller than the others, there is no significant performance alleviation for the entire simulation.

Algorithm 3 Pseudo-code for `SUBROUTINE edge-driven_level` (only for internal edges)

```
1:  !$omp simd simdlen(VL) aligned(∇Wx,W,zb,F,G,Sb :Nbyte)
2:  for i=1 ← [seg_y*(seg_x-1)] do
3:    j=i   ;   k=i+seg_y
4:    .........
5:    compute MUSCL_method + bed_slope + shallow_water_solver
          [∇Wx(j),∇Wx(k),W(j),W(k),zb(j),zb(k),...,Fx^L,Fx^R,Gx^L,Gx^R,Sbx^L,Sbx^R]
6:    F+G(i) ← Fx^L,Fx^R,Gx^L,Gx^R   ;   Sb(i) ← Sbx^L,Sbx^R
7:  end for
8:  for l=1 ← [seg_x] do
9:    m=seg_y*(seg_x-1)+1+(l-1)*(seg_y-1)   ;   n=m+seg_y-2   ;   o=(l-1)*seg_y
10:   !$omp simd simdlen(VL) aligned(∇Wy,W,zb,F,G,Sb :Nbyte)
11:   for i=m ← n do
12:     j=(i-m+1)+o   ;   k=j+1
13:     .........
14:     compute MUSCL_method + bed_slope + shallow_water_solver
            [∇Wy(j),∇Wy(k),W(j),W(k),zb(j),zb(k),...,Fy^L,Fy^R,Gy^L,Gy^R,Sby^L,Sby^R]
15:     F+G(i) ← Fy^L,Fy^R,Gy^L,Gy^R   ;   Sb(i) ← Sby^L,Sby^R
16:   end for
17: end for
```

Algorithm 4 Pseudo-code for `SUBROUTINE cell-driven level` (only for internal cells)

```
for k=1 ← [seg_x-2] do
  j = (seg_y+2)+(k-1)*seg_y   ;   l = (seg_y*(seg_x-1)+seg_y)+(k-1)*(seg_y-1)
  !$omp simd simdlen(VL) aligned(W,F,G,nm,Sb,Sf :Nbyte)
  for i=j ← [j+seg_y-3] do
    i1 = l+(i-j)   ;   i2 = i   ;   i3 = i1+1   ;   i4=i-seg_y
    .........
    compute friction_term [W(i),nm(i),...,Sf(i)]
    compute update_variables
    W(i) ← F+G(i1),F+G(i2),F+G(i3),F+G(i4),Sb(i1),Sb(i2),Sb(i3),Sb(i4),Sf(i)
  end for
end for
```

3.3. Avoiding Skipping Iteration for Vectorization of Wet–Dry Problems

In reality, almost all shallow flow simulations deal with wet–dry problems. To this end, the computations of both solver and bed-slope terms in the `SUBROUTINE edge-driven level` must satisfy the well-balanced and positivity-preserving properties as well, see [27,28], among others. Similarly, the calculations of the friction terms in the `SUBROUTINE cell-driven level` must also consider the wet–dry phenomena, otherwise errors are obtained. For example, in the edge-driven level, a wet–dry or dry–dry interface of an edge may exist since one or two cell-centers consist of no water; for both cases, the MUSCL method for achieving second-order accuracy is sometimes not required or even if this method is still computed, it must be turned back to first-order accuracy to ensure computational stability by simply defining the edge values according to the corresponding centers. Another example is in the cell-driven level, where the transformation of the unit discharges (hu and hv) back to the velocities (u and v) are required for computing the friction terms by a division of a water depth (h); very low water depth may thus cause significant errors. To anticipate these problems, one often employs some skipping iterations in the loops, see Algorithm 5.

Algorithm 5 Pseudo-code of some possible skipping iterations

```
!== This is a typical skipping iteration in the SUBROUTINE edge-driven level ==!
if [wet-dry or dry-dry interfaces at edges] then
  NO MUSCL_method: calculate first-order scheme
else
  compute MUSCL_method:  calculate second-order scheme
  if [velocities are not monotone] then
    back to first-order scheme
  end if
  .........
end if
!== This is a typical skipping iteration in the SUBROUTINE cell-driven level ==!
if [depths at cell-centers > depth limiter] then
  compute friction_term
else
  unit discharges and velocities are set to very small values
  .........
end if
```

Typically, the two skipping iterations in Algorithm 5 are important to ensure the correctness of shallow water models. Unfortunately, such layouts may destroy auto vectorization—or although a guided vectorization is possible, it does not give any significant improvement or may even decrease the performance significantly. This is because the SIMD instructions simultaneously work only for sets of arrays, which have contiguous positions. In our experiences, a guided vectorization was indeed possible for both iterations; the speed-up factors, however, were not so significant. Borrowing the idea of [22], we therefore change the layouts in Algorithm 5 to those in Algorithm 6, where the early exit condition is moved to the end of the algorithm. Using the new layouts in Algorithm 6, we significantly observed up to 48% more improvements of the vectorization from those given in Algorithm 5. Note that the results given by Algorithms 5 and 6 should be similar because no computational procedure is changed but only the layouts.

Algorithm 6 Pseudo-code of the solutions of the skipping iterations in Algorithm 5

```
!== A solution for the skipping iteration in the SUBROUTINE edge-driven level ==!
compute MUSCL_method:  calculate second-order scheme
.........
if [velocities are not monotone] then
  back to first-order scheme
end if
.........
if [wet-dry or dry-dry interfaces at edges] then
  NO MUSCL_method: calculate first-order scheme
end if
!== A solution for the skipping iteration in the SUBROUTINE cell-driven level ==!
compute friction_term
.........
if [depths at cell-centers ≤ depth limiter] then
  unit discharges and velocities are set to very small values
  .........
end if
```

3.4. Parallel Computation

We explain briefly here the parallel computing implementation of NUFSAW2D according to [21]. Our idea is to decompose and parallelize the domain based on its complexity level. NUFSAW2D employs hybrid MPI-OpenMP parallelization, thus is applicable to parallel simulations with multi-nodes. However, as we focus here on the vectorization, which no longer influences the scalability beyond one node [20], we limit our study on single-node implementations and thus only employ OpenMP for parallelization. Further, we examine the memory bandwidth effect when using only one core or 16 cores (AVX), 28 cores (AVX2), and 64 cores (AVX-512).

In Figure 4 we show an example of the decomposition of the domain in Figure 3 using four threads; for the sake of brevity, the illustration is given only for the edge-driven level. The parallel directive, e.g., `!$omp do`, can easily be added to each loop, thus according to Algorithm 2, in the gradient level the domain is decomposed as: thread 0 (cells 6, 7, 1, 17, 5, 8), thread 1 (cells 10, 11, 2, 18, 9, 12), thread 2 (cells 14, 15, 3, 19, 13, 16), and thread 3 (cells 4, 20). Similarly, regarding Algorithm 3 it gives in the edge-driven level: thread 0 (edges 1–4, 17–22, 32–33, 37–38, 42, 46), thread 1 (edges 5–8, 23–25, 34, 39, 43, 47), thread 2 (edges 9–12, 26–28, 35, 40, 44, 48), and thread 3 (edges 13–16, 29–31, 36, 41, 45, 49). Meanwhile, the cell-driven level applies a similar decomposition to that of the gradient level. One can see, the largest loop components, e.g., internal edges 1–4, 5–8, etc., are decomposed in a contiguous pattern easing the vectorization implementation, thus efficient. Note the decomposition in Figure 4 is based on static load balancing that causes load imbalance due to the non-uniform amount of loads assigned to each thread; this load imbalance will become less and less significant as the domain size increases, e.g., to millions of cells. However, another load imbalance issue—which can only be recognized during runtime—appears, namely the one caused by wet–dry problems, where wet cells are computationally more expensive than dry cells. For this, we have developed in [21] a novel weighted-dynamic load balancing (WDLB) technique that was proven effective to tackle load imbalance due to wet–dry problems. All the parallel and load balancing implementations are described in detail in [21], thus are not explained here. We also note that we have successfully applied this cell-edge reordering strategy in [24,25] for parallelizing the 2D shallow flow simulations using the CU scheme with good scalability. Yet, we will show in the next section that the cell-edge reordering strategy proposed can help in easing all the vectorization implementations.

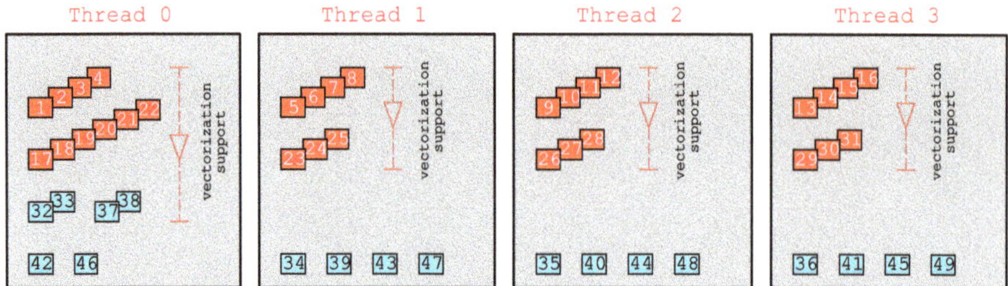

Figure 4. Illustration of load distribution using static load balancing with vectorization support for the edge-driven level based on Figure 3.

4. Results and Discussions

We validate our model against four benchmark tests: two dam-break cases and two tsunami cases. Each case was simulated using a constant Δt that satisfies the Courant-Friedrichs-Lewy (CFL) condition, where CFL $\leq$ 0.5. Our model with the HLLC, Roe, or CU scheme satisfies the well-balanced property; also, the HLLC and CU solvers employed are positivity-preserving. We note that the Roe scheme may in some cases produce negative depths, see [29]; however, in all implementations tested here, we did not find any negative depth with the Roe scheme. The Δt used also fulfills the CFL limitation required by the computations of the local one-sided propagation speeds of the CU scheme for positivity-preserving purpose, see [13].

4.1. Case 1: Circular Dam-Break

This case is included to check the capability of our model for symmetry and shock resolution in shallow water flow modeling. We refer to [16,30], among others. A 40 $\times$ 40 m, flat, and frictionless domain is considered. A cylindrical wall with a radius of 2.5 m, which was centered at the domain, separated two regions of still water; the first one inside the cylinder had a depth of 2.5 m and the second one outside consisted of 0.5 m water. The water was assumed to be initially at rest and all boundaries were set to wall boundary. The main features to be investigated in this case are the rarefaction wave and the hydraulic jump (shock wave) including a transition condition from subcritical to supercritical flow. The total simulation time was set to 4.7 s with Δt = 0.005 s, thus requiring 940 time steps. The domain was discretized into 160,000 rectangular cells (319,200 edges).

The evolutions of the simulated free surface elevation using the CU scheme are visualized in Figure 5. Suddenly after 0.1 s, water started to move in all directions. At 0.4 s, the circular shock wave propagated outwards, whereas the circular rarefaction wave traveled inwards showing that this wave almost reaches the center of the domain. This phenomenon continued until the rarefaction wave has fully plunged into the center of the domain at approximately 0.8 s and this wave was suddenly reflected creating a sharp gradient of water surface elevation. At 1.6 s, the circular shock wave propagated further outwards the from domain center, whereas the reflected rarefaction wave now caused the water to fall below the initial depth of 0.5 m. This produced a secondary circular shock wave, the depth of which was slightly less than 0.5 m. The primary circular shock wave kept propagating outwards the center of the domain at 3.8 s and interestingly, the secondary circular shock wave that had recently been created traveled towards that center. At 4.7 s, it is shown that the primary circular wave almost reached the domain boundary and at this time a very sharp gradient of water surface elevation had been created near that boundary.

We present the comparison between the analytical and numerical results at 4.7 s in Figure 6 showing that all schemes can simulate this highly discontinuous flow properly. To point out the difference between the three schemes more clearly, we present in Figure 7 both the depth and velocity profiles near the two discontinuous areas: 20–22 m and 38–40 m, where only non-significant differences are shown.

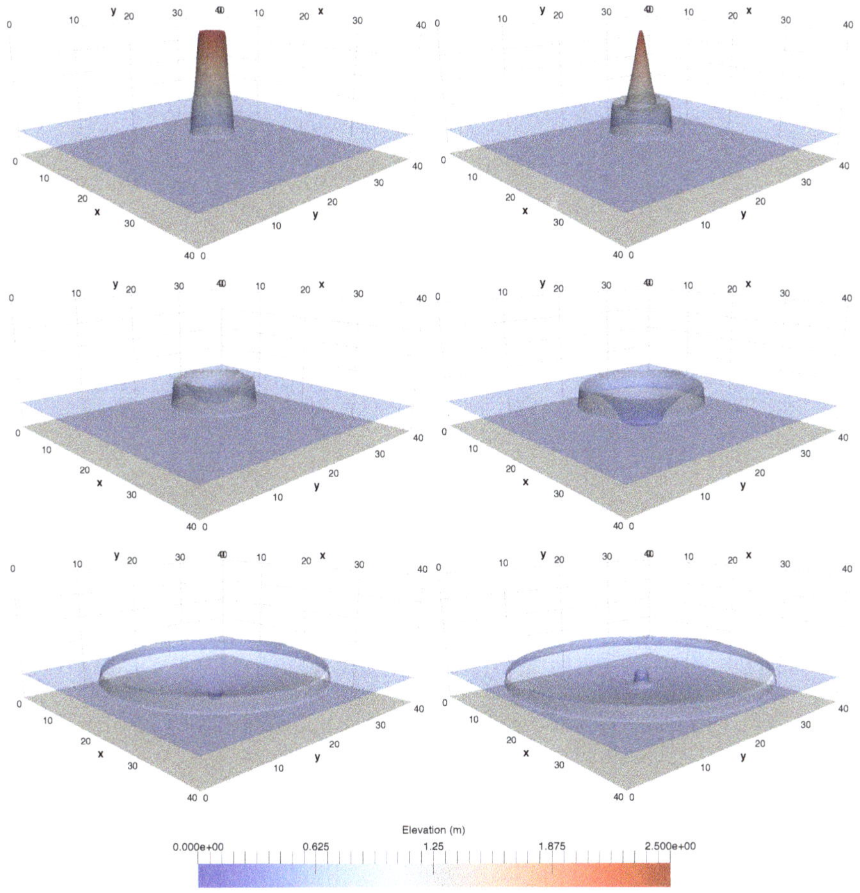

Figure 5. Case 1: results of the central-upwind (CU) scheme at 0.1, 0.4, 0.8, 1.6, 3.8, and 4.7 s.

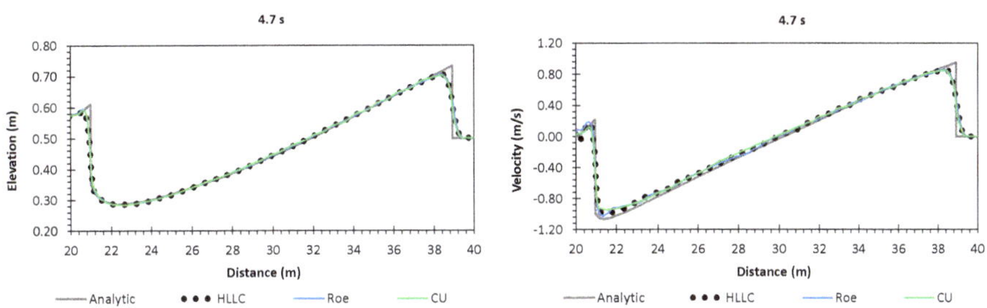

Figure 6. Case 1: comparison between analytical and numerical results at 4.7 s.

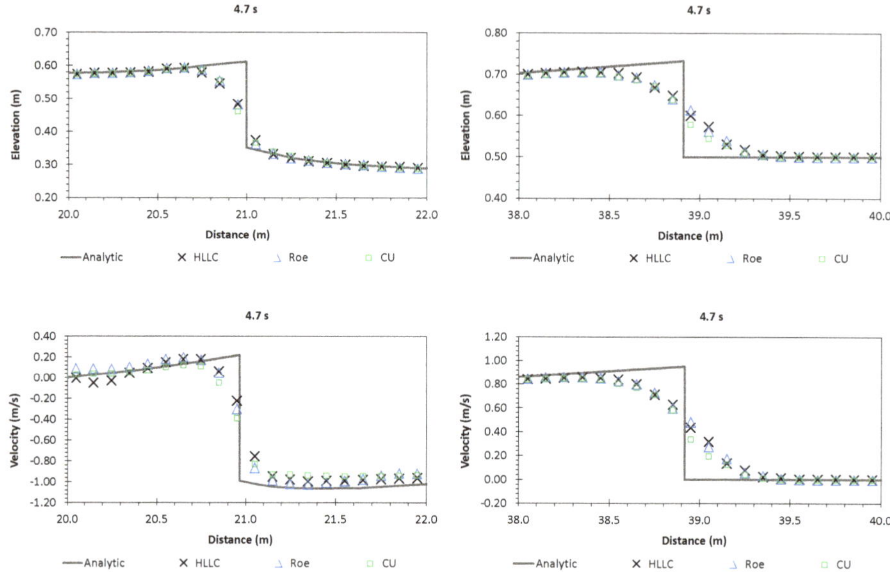

Figure 7. Case 1: comparison between analytical and numerical results at 4.7 s (in detail).

4.2. Case 2: Dam-Break Flow against an Isolated Obstacle

This case was done experimentally in [31]. The channel was trapezoidal; 35.8 m long and 3.6 m wide. A 1 m wide rectangular gate separated the upstream reservoir from the downstream channel, see Figure 8. The Manning coefficient was 0.01 s m$^{-1/3}$. A 0.8 $\times$ 0.4 m obstacle was located on the downstream channel with a position that formed an angle of 64° from the x-axis. The water was set initially to 0.4 m at the reservoir and 0.02 m at the channel, thus the banks at downstream were dry. The upstream end of the reservoir was a closed wall. In this paper, the domain was discretized into 143,280 rectangular cells (285,246 edges). The simulation was set for 30 s with Δt = 0.005 s, thus requiring 6000 time steps.

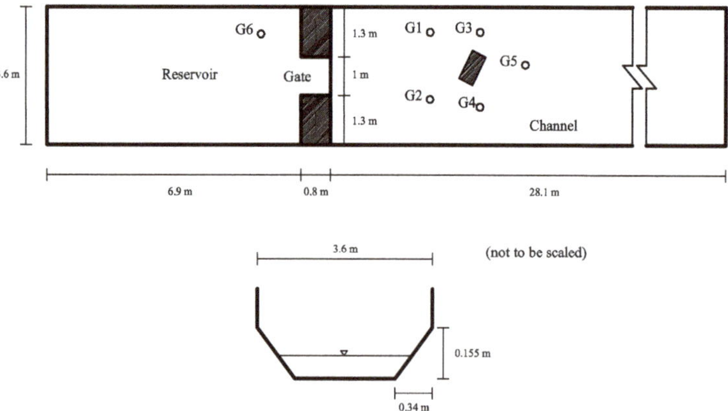

Figure 8. Case 2: sketch of domain and channel shape.

We compared our model at four points: G1 (10.35, 2.95) m, G4 (11.7, 1.0) m, G5 (12.9, 2.1) m, and G6 (5.83, 2.9) m. Our numerical results are given in Figure 9 showing that our model is in general capable of simulating this case properly. At G1, the maximum bore around 2 s was accurately simulated by all schemes, where there were no significant differences shown until 9 s. However, after 9 s, the CU scheme computed the results higher than do the other schemes, where both the HLLC and Roe schemes show almost no different results. At G4, the first bore around 2 s was predicted with a later time of no more than 1 s and a higher depth of no more than 2 cm, where all schemes kept producing the higher values from 2 s to 4.5 s. At G5, no significant differences were again shown between the HLLC and Roe schemes, but the CU scheme showed slightly different values. At G6, highly accurate results were given by all schemes to simulate the water at the reservoir, showing that the schemes can predict the correct incoming discharge from the upstream reservoir to the downstream channel.

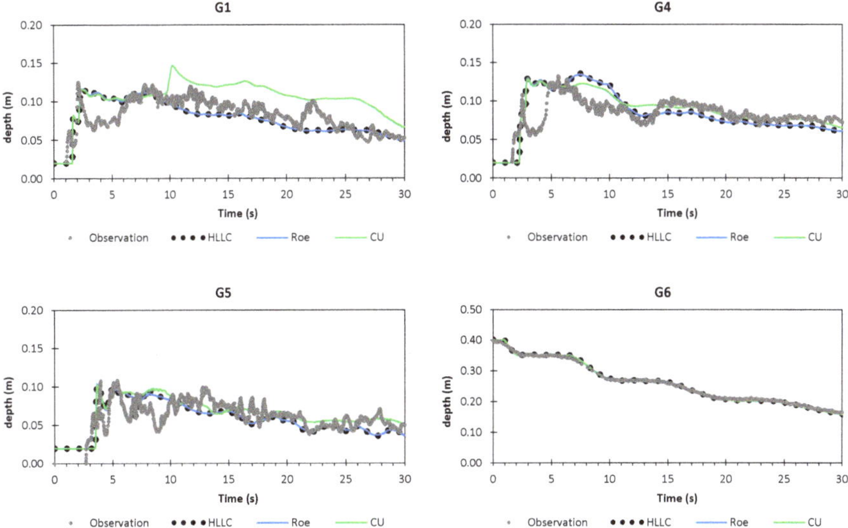

Figure 9. Case 2: comparison of depths between observation and numerical results.

Some errors computed by our model are probably due to the absence of the turbulence terms. Yu and Duan [32] showed the turbulence model was highly important for simulating flow field around the obstacle, where the reflection waves from the obstacle and side walls have superimposed several oblique hydraulic jumps. In Figure 10, we visualize the flood propagation at 1, 3, and 10 s using the CU scheme.

4.3. Case 3: Tsunami Run-Up on a Conical Island

This benchmark case was conducted in a laboratory by [33] to investigate the tsunami run-up on a conical island, the center of which was located near the middle of a 30 × 25 m basin, see Figure 11. To produce planar solitary waves with the specified crest and length, a directional wave maker was used. The left boundary was set as a flow boundary, and the respective water elevation and velocities were defined as

$$\begin{aligned} \eta(0,y,t) &= A_e \,\text{sech}^2 \sqrt{\frac{3\,A_e}{4\,H_e}} \sqrt{g\,\left(H_e + A_e\right)}\,\left(t - T_e\right)\,, \\ u(0,y,t) &= \frac{\eta\,\sqrt{g\,\left(H_e + A_e\right)}}{\eta + H_e}\,, \qquad v(0,y,t) = 0\,, \end{aligned} \tag{7}$$

where A_e, H_e, and T_e are the amplitude of the incident wave, still water depth, and time, at which the wave crest enters the domain—set to 0.032 m, 0.32 m, and 2.45 s, respectively. The other three boundaries were closed boundaries. We compared our results with the values at five gauges located on the domain: P-03, P-06, P-09, P-16, and P-22, whose coordinates were (6.82,13.05) m, (9.36, 13.80) m, (10.36, 13.80) m, (12.96, 11.22) m, and (15.56, 13.80) m, respectively. The Manning coefficient was set to zero as suggested by [34].

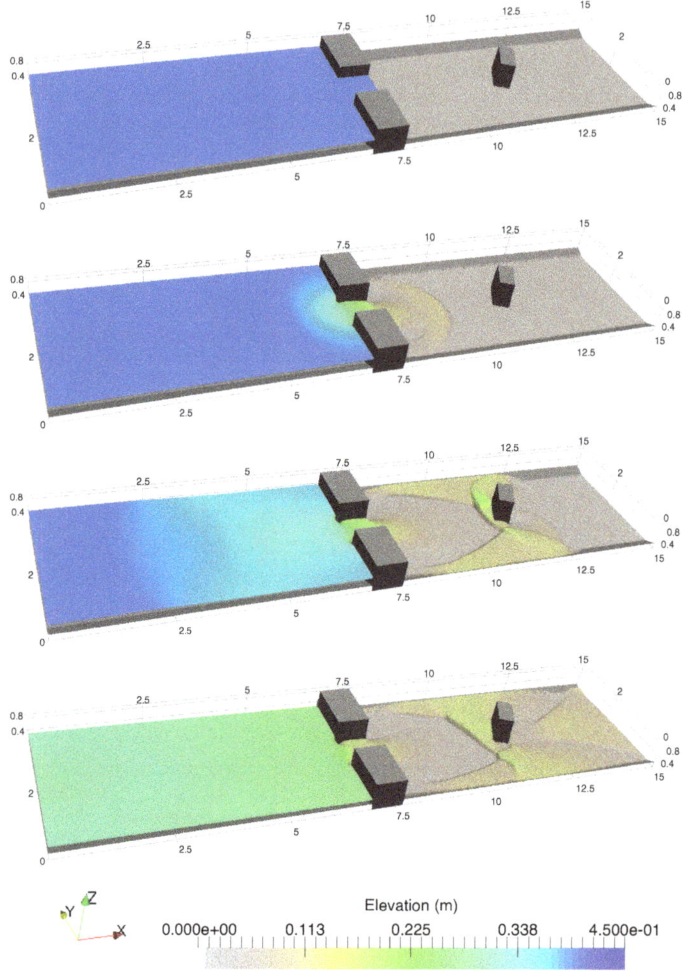

Figure 10. Case 2: visualization of the flood propagation at 0, 1, 3, and 10 s using the CU scheme.

The domain was discretized into 200,704 rectangular cells (402,304 edges). The simulation time was set to 20 s with $\Delta t = 0.002$ s leading to 10,000 time steps. One can see in Figure 12, the incident solitary waves in front of the island, which generate a high run-up at about $t = 9$ s, create wet–dry mechanisms on the conical island. Within this period, the maximum magnitude was reached. After $t = 9$ s, the waves started to run down the inundated area on the conical island. Some waves were refracted and propagated toward the lee side of the island, where two waves were trapped at each side of the island at around $t = 11$ s. At $t = 13$ s, the second wave run-up was generated after these two waves collided. Afterwards, these waves continued to propagate around the island.

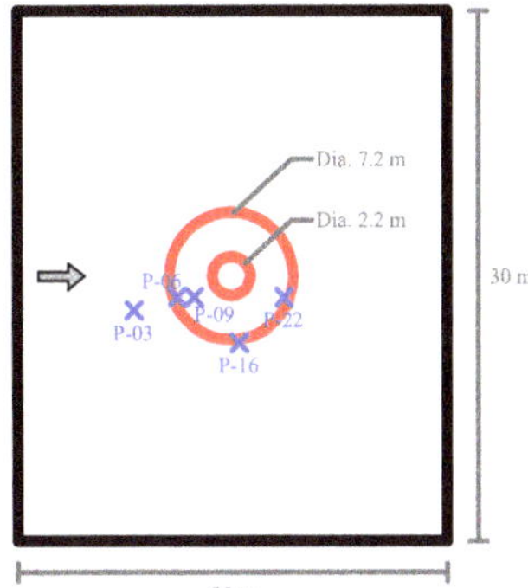

Figure 11. Case 3: computational domain of solitary wave run-up.

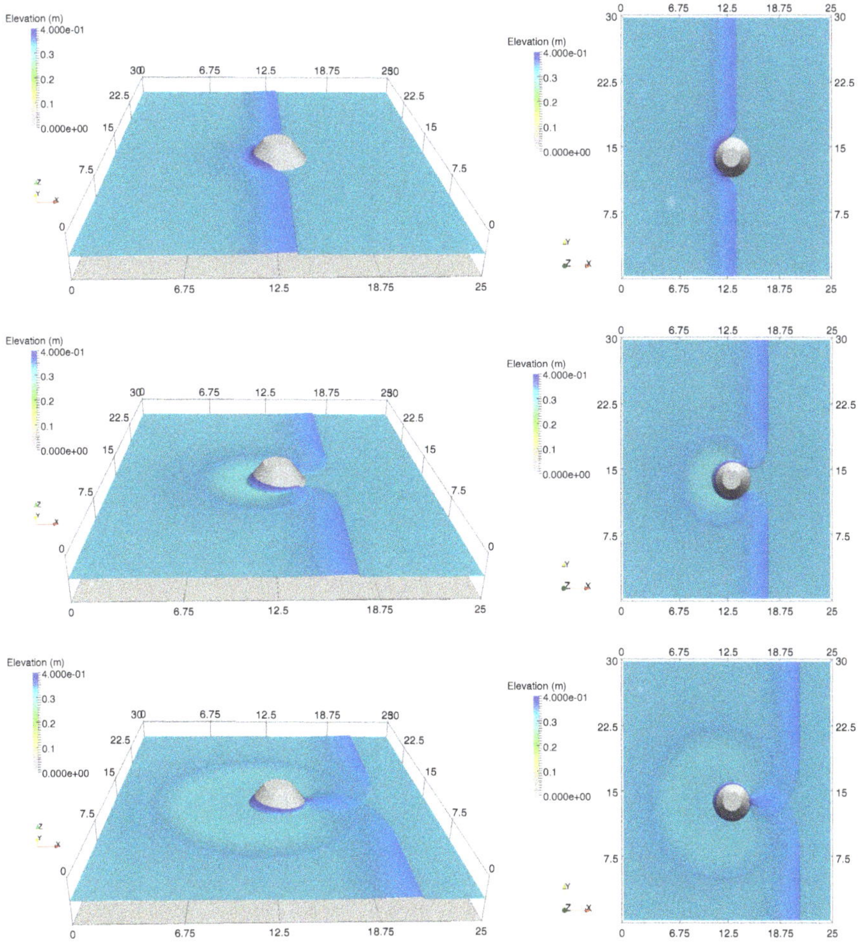

Figure 12. Case 3: numerical results using the CU scheme at 9, 11, and 13 s.

Our numerical results are also compared with laboratory results during 20 s, see Figure 13. Accurate results were produced by all schemes, where no significant differences between them were shown. The arrival times of the highest waves were accurately detected at gauges P-03 and P-22. All schemes rendered later times at gauges P-06, P-09, and P-16 but the differences were no more than 1 s. At gauge P-16, our model computed the wave 1 cm higher than the one mentioned in the laboratory data, and the wave at gauge P-22 was computed 1.3 cm higher. This was probably due to the neglect of the dispersion effects. Note that such discrepancies were also reported in the numerical model of [34].

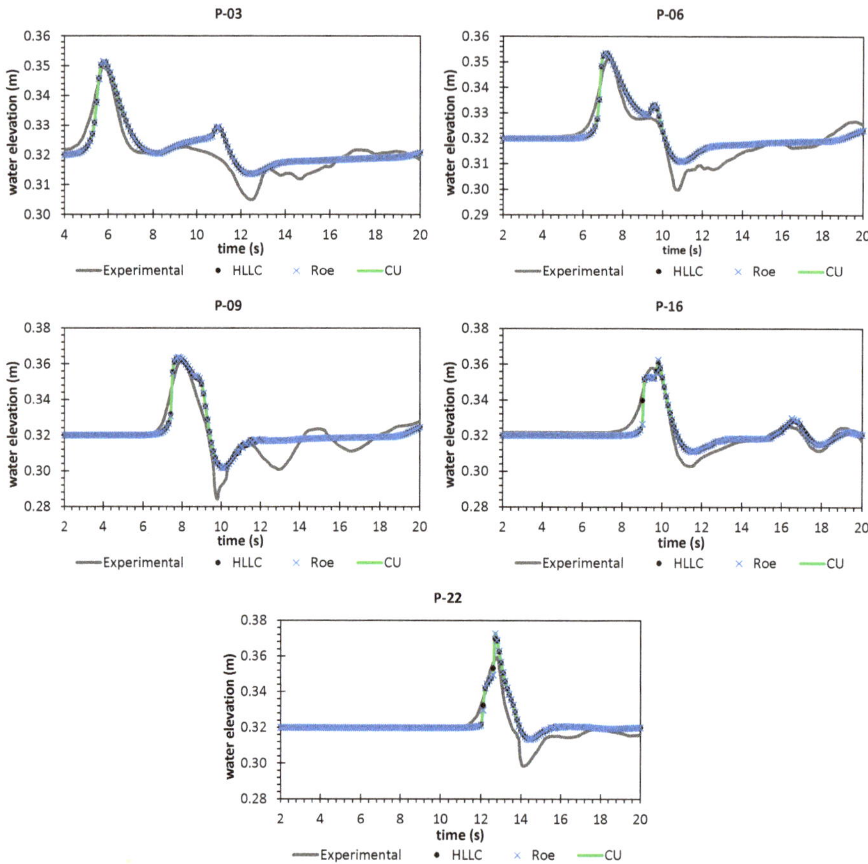

Figure 13. Case 3: comparison between observation and numerical results.

4.4. Case 4: 2011 Japan Tsunami Recorded in Hawaii

This benchmark test is a real tsunami case that occurred in 2011, Japan. The data set was recorded in Hilo Harbor, Hawaii. The raw data can be found in [35]. To avoid the phase differences of the incident wave, the original bathymetry data should be flattened at the depth of 30 m. Interested readers are also referred to [36] for more information. In Figure 14, the sketch of the domain is given as well as the incident wave forcing employed at the northern part as a boundary condition. The Manning coefficient was assumed to be uniform 0.025 s $m^{-1/3}$. The observation points were the Hilo tide station for elevation (3159, 3472) m, HAI1125/harbor entrance (4686, 2246) m, and HAI1126/inside harbor (1906, 3875) m for velocities. The 1-minute de-tiding of raw data was done for the observed data.

The domain was discretized using 20 m resolution rectangular cells producing 94,600 rectangular cells (189,200 edges). We set the simulation time to 13 h and used $\Delta t = 0.025$ s giving 1,872,000 time steps. The results are given in Figure 15 plotted per 150 s. At the Hilo tide Station, each scheme can detect the first incoming wave quite accurately around $t = 8.2$ h. The lowest water elevation was also predicted properly at approximately t = 8.4 h but with a non-significant difference of about 0.2 m. After that, the water level fluctuations were also computed properly. At the harbor entrance, the velocities were in general accurately computed. Each scheme was able to compute the first incoming wave for the x velocity at $t = 8.2$ h. The y velocity magnitude at that time was, however, slightly overestimated. Inside the harbor, accurate predictions for x and y velocities were shown, where the first incoming wave was well predicted. After 10 h, each scheme kept exhibiting accurate results at the harbor entrance as well as inside the harbor. One can see that the water current flowed predominantly in North–South direction at the harbor entrance, whereas inside the harbor the water current flowed predominantly in East–West direction. Our results agree with the observed data and those simulated by [36] as well. Although some discrepancies—which are probably due to the neglect of the tidal current effects—still exist, our model shows overall quite accurate results for this hazard event.

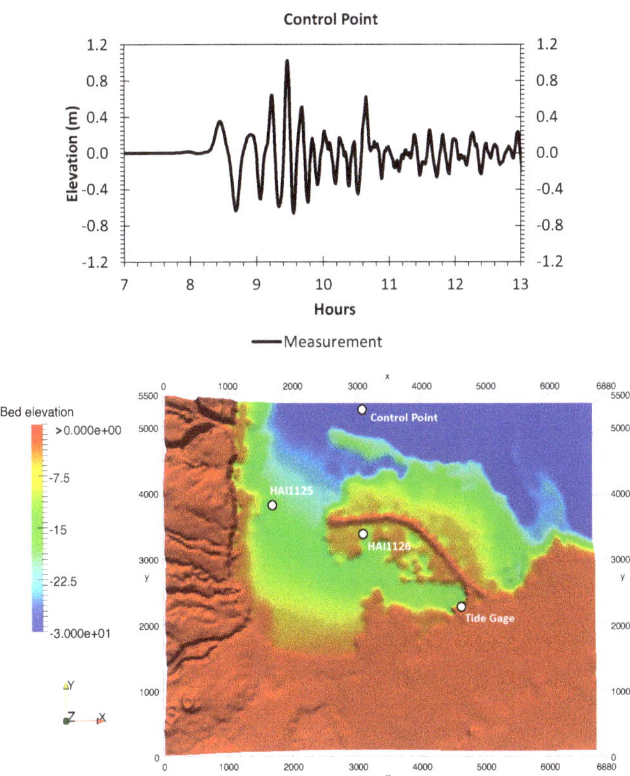

Figure 14. Case 4: bathymetry for simulation and the boundary condition.

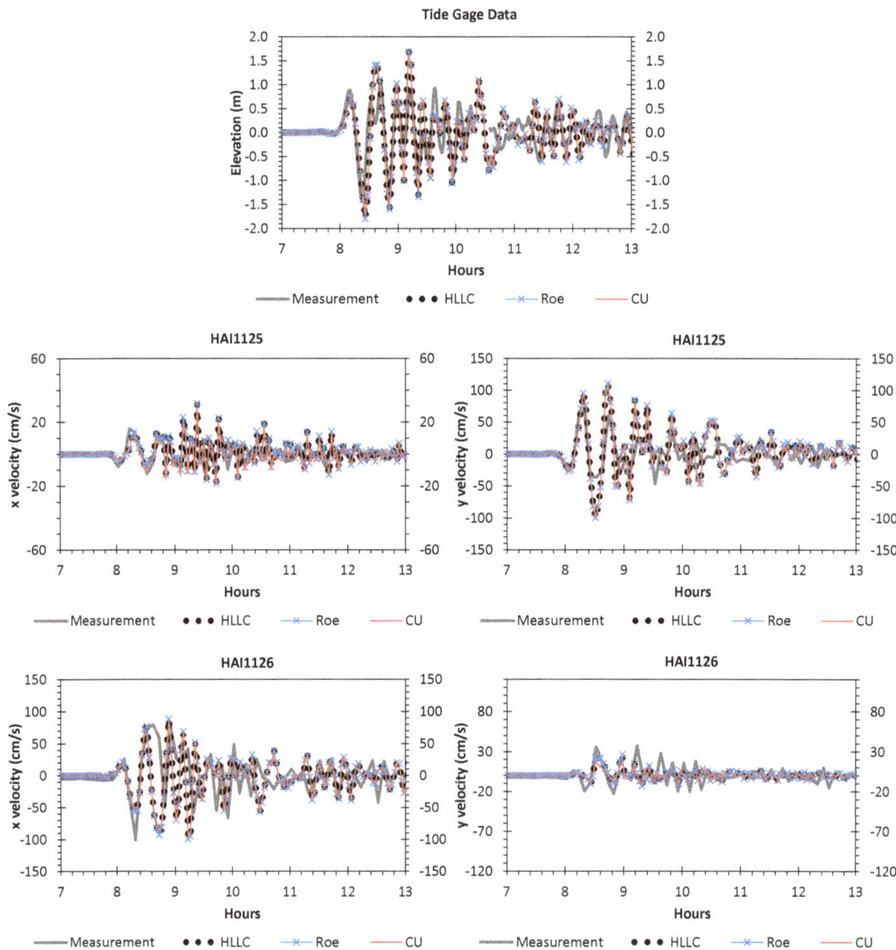

Figure 15. Case 4: comparison between observation and numerical results.

In Figure 16, the visualizations of tsunami inundation are presented using the CU scheme. It is shown at around 9.03 h, the water level reaches approximately 0.5–1 m at the harbor entrance. Meanwhile, the water level is predicted to reach 1–1.5 m inside the harbor. At about 9.53 h, the water level at the harbor entrance remains relatively constant for 0.5–1 m but outside the harbor (near the breakwater) the water level becomes higher up to 2.5 m. After 14 h, the water level near the breakwater (inside and outside the harbor) decreases to approximately -1.25 m. Complex wet–dry phenomena near the coastline as well as the breakwater appear during the simulation time and our model has shown to be robust for modeling such phenomena.

We show in Figure 17 a visualization of the maximum velocity magnitude captured by the HLLC, Roe, and CU schemes during 13 h simulation time. In general, as one can see, no significant differences are shown between all schemes. Along the outer side of the breakwater as well as near the harbor entrance, the velocity magnitudes of more than 4.5 m/s appear. Meanwhile, considerably lower magnitudes are shown inside the harbor. The main difference is only located near the harbor entrance, where the CU scheme computes the slightly lower magnitudes. The spatial distribution of the velocity magnitude is shown to be extremely sensitive, in agreement with that studied in [36].

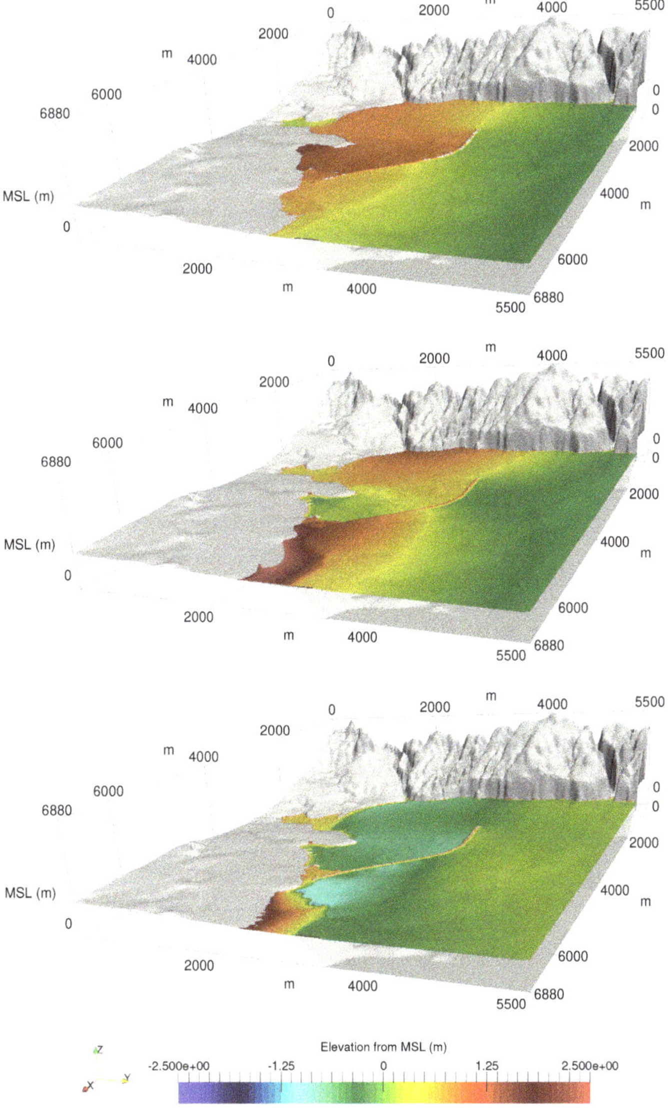

Figure 16. Case 4: visualization of tsunami inundation using the CU scheme at 9.03, 9.53, and 11 h.

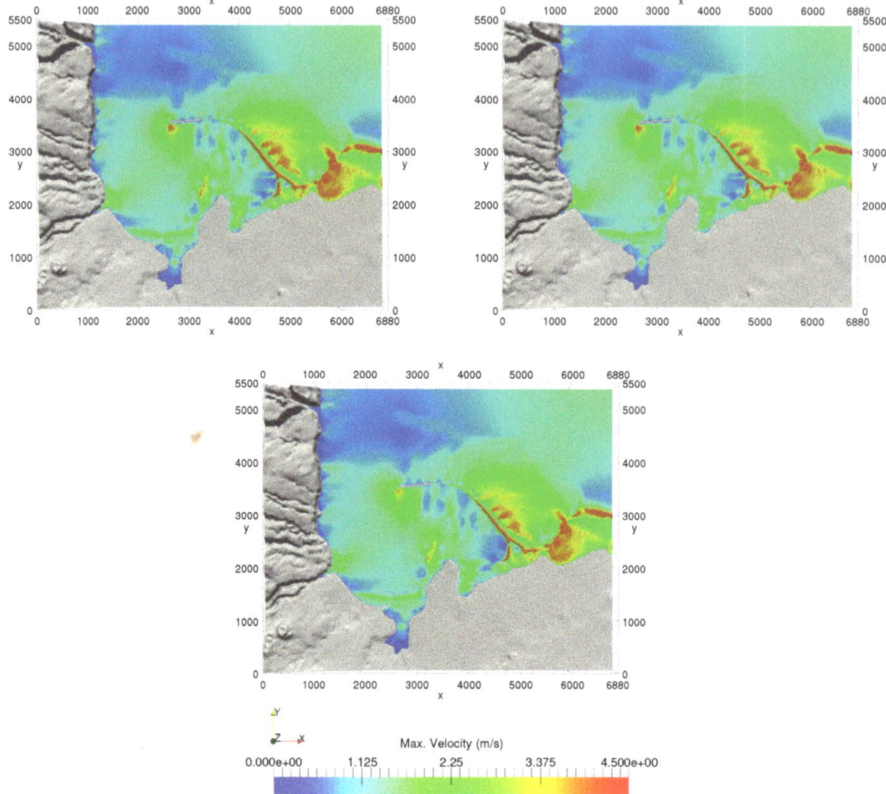

Figure 17. Case 4: numerical result for the maximum velocity captured during the 13 h simulation time using the Harten-Lax-van Leer-Contact (HLLC), Roe, and CU schemes (top left, top right, bottom).

4.5. Performance Comparison

We have shown in the previous sections that the HLLC, Roe, and CU schemes are quite accurate for simulating the test cases, where only non-significant differences are shown between them. In this section we analyze and compare the performance of each scheme. All schemes were written and compiled in the same code NUFSAW2D on three machines—AVX (Intel Xeon E5-2690/Sandy-Bridge-E), AVX2 (Intel Xeon E5-2697 v3/Haswell), and AVX-512 (Intel Xeon Phi/Knights Landing)—for a Linux operating system using Intel Fortran 19. The first computing resource "Sandstorm" was available at our chair [37] and the last two resources "CoolMUC-2" and "CoolMUC-3" were provided by the Leibniz Supercomputing Centre (LRZ) [38]. Each node of the AVX, AVX2, and AVX-512 machines has a total of eight physical cores (16 logical cores), 14 physical cores (28 logical cores), and 64 physical cores, respectively. Note that AVX-512 is built on many-core architecture that incorporates cores with low-frequency and small memory. Therefore, in order to achieve a notable performance, this machine relies on the vector operations on 512 bit SIMD registers.

We did not use the vectorization directive provided by Intel, e.g., `!dir$ simd`, since we have experienced that this directive was not always able to vectorize the loop. Instead, we implemented the directive `!$omp simd simdlen(VL) aligned(var1,var2,... :Nbyte)` provided by the OpenMP 4.0. The first component (`simdlen`) was aimed to test the benefit of vectorization on our code compared to the theoretical speed-up based on the vector width, while the second one (`aligned`) was employed to know the benefit of the aligned memory accesses supported by the reordering strategy proposed. Since we would like to emphasize the effect of vector width, we restricted our discussion here to single-precision arithmetic. The variable `VL` was the vector length, set to eight for AVX/AVX2 and 16 for AVX-512—and `Nbyte` was the default alignment of the architecture, set to 32 for AVX/AVX2 and 64 for AVX-512.

Two metrics are used to denote the performance of our code: Medge/s/core (million edges per second per core) and Mcell/s/core (million cells per second per core), which are the comparisons between the total number of simulated edges or cells that can be achieved per unit of time using one core. The former was used to denote the performance of the `SUBROUTINE edge-driven level`, whereas the latter was used to denote the performance of the entire simulation. It is also important to note that since the RKFO method is used, the latter is calculated after four times updating per time-level update, not per calculation-level update. We compiled our code using the flag `-O3 -qopenmp -align 'a' 'b'` for the vectorized version, where `'a' = array32byte` for AVX/AVX2 and `array64byte` for AVX-512—and `'b' = -xAVX` for AVX, `-xCORE-AVX2` for AVX2, and `-xCOMMON-AVX512` for AVX-512. To only emphasize the performance increase by vectorization, we disable all possibilities for auto vectorization by compiling with the flag `-O3 -qopenmp -no-vec -align 'a' 'b'` and by deleting all the SIMD directives in the source code, thus giving a fair benchmark of the non-vectorized version of our code.

As previously explained, we discuss our results using single-core and single-node computations. We observed that for single-node computations, NUFSAW2D with OpenMP gives better performance than MPI because the WDLB technique employed for wet–dry problems requires no communication cost. We only performed strong scaling for all cases, where we achieved averagely 87% efficiency for AVX/AVX2 with 16/28 cores and 88% efficiency for AVX-512 with 64 cores. When using 8/16 cores with AVX/AVX2 or 56 cores with AVX-512, higher efficiency was even achieved by our code being approximately 98%. Although this leads to a better performance, we still use the results with all cores available to show the single-node performance. Note the performance degradation of 12–13% (when using all cores) was not due to inefficient load distribution but probably because of the non-uniform memory access (NUMA) effects, where a processor can access its memory faster than the shared non-local memory, see [21].

4.5.1. Performance of Edge-Driven Level

Figure 18 shows the performance comparison between all solvers, in which we observe a significant performance improvement for each solver. Note the results in Figure 18 represent the average values from the four cases tested. We observed that there are no significant differences of the performance (in the range of 4–5%) achieved in all cases. The worst performance was shown in case 2, whereas the best one was achieved in case 1. This is because case 2 deals with more complex wet–dry problems, for which the WDLB technique in this case works better than in the other cases—thus causing more overheads—in order to balance the load units between wet and dry cells, see [21] for detail. For the edge-driven level, each non-vectorized solver shows performance metrics with a range of 3.42–4.54, 5.03–6.23, and 1.01–1.38 Medge/s/core for the AVX, AVX2, and AVX-512, respectively. This shows the CU scheme was, without vectorization, averagely 1.31× and 1.26× faster than the HLLC and Roe solvers, respectively.

As soon the guided vectorization was activated, the performances of each scheme in the edge-driven level increased significantly. For the AVX machine (1 core), we observed significant improvements being 5.5×, 6.5×, and 6× for the HLLC, Roe, and CU schemes, respectively; this shows the Roe scheme experiences remarkably the benefit of the vectorization, for which the improvement factor is larger than the others. For the AVX2 machine (1 core), the speed-up factors of 4.5×, 4.8×, and 5× were obtained by the HLLC, Roe, and CU schemes, respectively showing that the improvement factor of the CU scheme becomes the largest one among the others. Although significant performance improvements have been shown, our model still cannot fully exploit the theoretical speed-up of 8×

from the vector widths of both the AVX and AVX2 machines used. Nevertheless, we have shown that the data structures of our code are suitable for SIMD instructions as we are able to achieve the efficiency of up to 81.25%. Note in the aforementioned notable works, none of the models could achieve the performance increase of more than 52% from the theoretical speed-up of the machine used. In [20], the average speed-up of 4.1× was achieved on AVX machine (single-precision) for the vectorized augmented Riemann solver; therefore, this leads to the efficiency of 51.25%. In [22], the average speed-up of 1.7× was obtained on an AVX2 machine (double-precision) for the vectorized Riemann solver; this thus gives the efficiency of 42.5%.

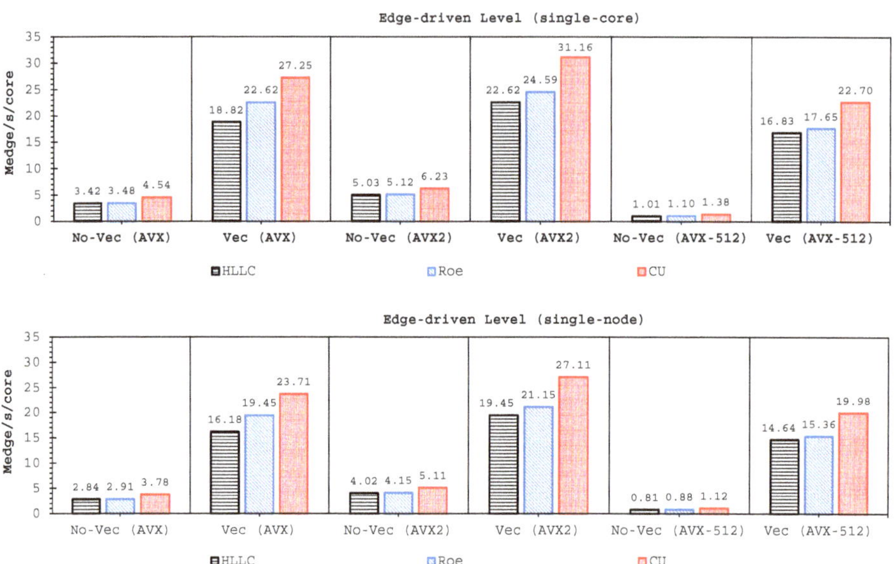

Figure 18. Comparison of performance metrics in the edge-driven level.

For the AVX-512 machine (1 core), the vectorization has tremendously increased the performances of the HLLC, Roe, and CU solvers by the factors of 16.68×, 16.04×, and 16.42×, respectively. This shows that our model can comprehensively exploit the vectorization for the vector width provided, of which the theoretical speed-up is 16×. Also, this represents that the data structures designed in NUFSAW2D efficiently support the vector programming on this vector-computing architecture.

With parallel simulations, each non-vectorized solver exhibits the performance metrics within the ranges of 2.84–3.78, 4.02–5.11, and 0.81–1.12 Medge/s/core for the AVX (16 cores), AVX2 (28 cores), and AVX-512 (64 cores), respectively; compared to the non-vectorized values with 1 core, it gives about 83% efficiency. For the performance analysis of the parallelized-vectorized solvers, the values obtained by the non-vectorized solvers with single-core are used as indicator. For the AVX machine (16 cores), the parallelized-vectorized HLLC, Roe, and CU solvers reached 16.18, 19.45, and 23.71 Medge/s/core, giving speed-ups of 75.7×, 89.4×, and 83.52×, respectively. Similarly, the parallelized-vectorized HLLC, Roe, and CU solvers obtained 19.45, 21.15, and 27.11 Medge/s/core, respectively with the AVX2 machine (28 cores) leading to speed-ups of 108.4×, 115.6×, and 121.8×. The significant performance increase was shown by the AVX-512 machine (64 cores), where the parallelized-vectorized HLLC, Roe, and CU solvers reached 14.64, 15.36, and 19.98 Medge/s/core, respectively; this brings each scheme to achieve speed-ups of 928.9×, 892.9×, and 924.7×.

The results in Figure 18 show an interesting fact, especially for the single-node performance analysis. Without vectorization, the parallelized results of the AVX2 machine can significantly outperform the parallelized results of the AVX-512 machine. For example, see Figure 19, on the AVX2 machine the CU scheme shows a metric of 143.1 Medge/s with 28 cores while on the AVX-512 machine with 64 cores this scheme exhibits a metric of 71.7 Medge/s; the difference is thus almost two-fold. However, with vectorization, the parallelized results of the AVX2 machine (759.1 Medge/s) are now outperformed by those of the AVX-512 machine (1278.6 Medge/s), being approximately 1.7×. This shows the vectorization is non-trivial for increasing the performance.

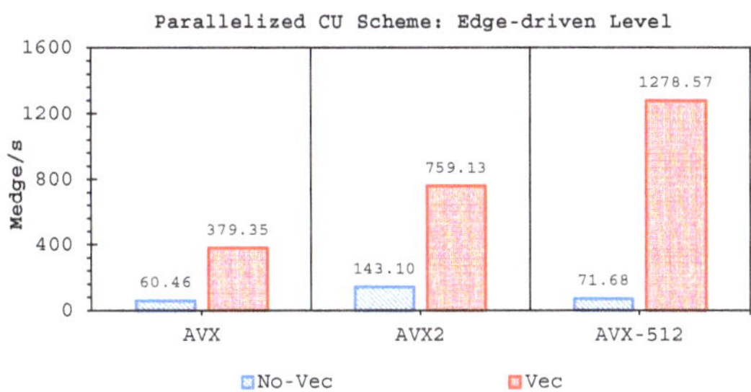

Figure 19. Single-node performance of the non-vectorized and vectorized CU scheme with parallelization (edge-driven level).

Based on these results, one can see although the Roe scheme experiences the largest speed-up on the AVX machine or the HLLC scheme achieves the largest improvement factor on the AVX-512 machine, both of these schemes are still significantly outperformed by the CU scheme with average multiplication factors of 1.4× and 1.25×, respectively. This is not so surprising since the computational procedures of both the HLLC and Roe solvers include complex branch statements (`if-then-else`), thus should theoretically be much more expensive than the CU scheme, see [17]. The HLLC scheme requires the nested branch statements; the first one is to compute the wave speeds, which are later required in the second branch statement for calculating the final convective fluxes. The Roe scheme needs branchings for the intermediate variables and entropy correction computations, the computations of which are quite complex. Such branchings may force the uses of masked operations and assignments, thus significantly decreasing the performance. In contrast to these two solvers, the CU scheme does not experience any branch statement. This is the beauty of this scheme in addition to being quite simple and having no complex procedure, thus can (even) be auto-vectorized by the compiler.

4.5.2. Performance of the Entire Simulation

Prior to investigating the performance of the entire simulation, we firstly show the cost estimation of each level in Algorithm 1 by presenting in Figure 20 a list of cost percentages: initialization, gradient, edge-driven level and cell-driven level. The last three components indicate the same levels to those shown in Algorithm 1, while initialization is a part required for updating the initial value for the RKFO method per time-level update, e.g., to perform $\mathbf{W}^{p=0} = \mathbf{W}^{t}$, see Equation (1). Note for an unbiased representative, the values in Figure 20 are the cost percentage of a vectorized solver relatively to its non-vectorized version. Only the cost percentage of the simulations using single-core is presented in Figure 20; the percentage for single-node is shown to be similar. As expected, we observe that the edge-driven level is the most time-consuming part being 65–75% of the entire simulation for the non-vectorized code. For both AVX and AVX2 machines, the vectorization can decrease the computational cost of the edge-driven level approximately from 71% up to 15%. Meanwhile, for the AVX-512 machine, the vectorization is shown more effective to reduce the cost of the edge-driven level averagely from 72% up to 5%.

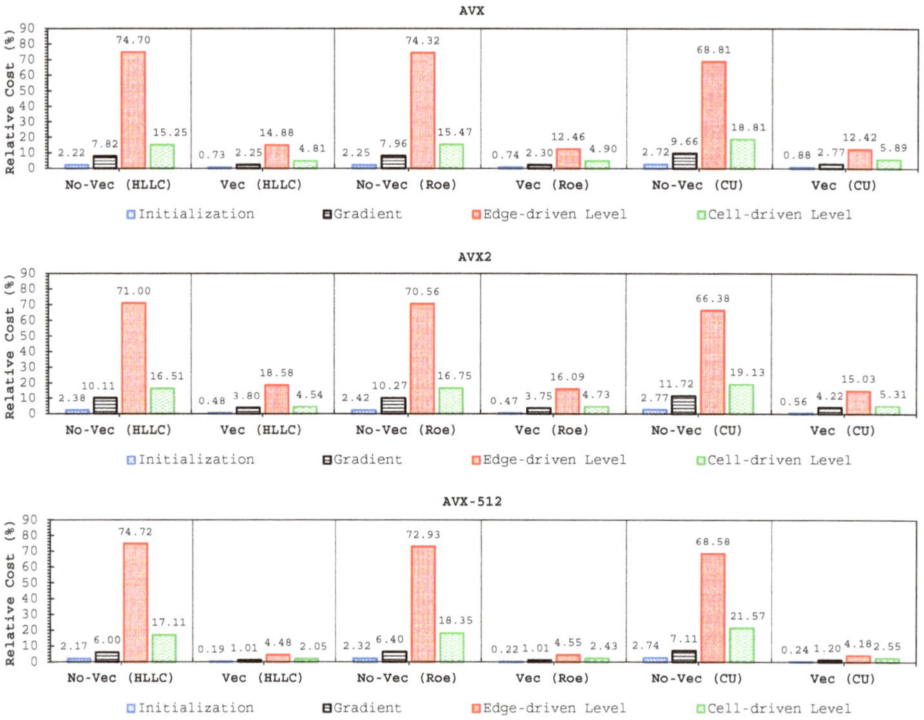

Figure 20. Components of the entire simulations for all schemes.

The second most expensive part is the cell-driven level, which consumes around 16–22% of the total simulation time with the non-vectorized solvers. After vectorization, the cost of the cell-driven level decreases from approximately 17% up to 5% on both the AVX and AVX2 machines. Meanwhile, on the AVX-512 machine, the vectorization has helped by decreasing the computational time of the cell-driven level averagely from 19% up to 3%; this again shows the vectorization works more effectively on this machine.

We now explain the performance of our model for updating the entire simulation. For the AVX, AVX2, and AVX-512 machines with one core, we observed for the non-vectorized solvers the metrics of 1.27–1.56, 1.78–2.06, and 0.38–0.47 Mcell/s/core, respectively—and for the vectorized solvers by 6.37–7.79, 7.12–9.28, and 5.23–6.23 Mcell/s/core, respectively. We achieved the improvements for the AVX machine by 5×, 5.5×, and 5× for the HLLC, Roe, and CU schemes, respectively, while for the AVX2 machine, the speed-up factors of 4×, 4.5×, and 4.5× were obtained by the HLLC, Roe, and CU schemes, respectively. On the AVX-512 machine we observed the speed-up factors of 13.91×, 13.11×, and 13.18× for the HLLC, Roe, and CU schemes, respectively showing that, on this machine, our code can achieve a better performance than those on the other two machines. However, the AVX2 machine still gives the highest metrics among the others.

For parallel simulations with the AVX, AVX2, and AVX-512 machines, the non-vectorized solvers achieved the metrics of 1.05–1.28, 1.42–1.67, and 0.3–0.38 Mcell/s/core, respectively—and for the parallelized-vectorized solvers by 5.48–6.78, 6.12–8.08, and 4.55–5.48 Mcell/s/core, respectively. Similar to the previous analysis, the values obtained by the non-vectorized solvers with single-core are used as indicator here. According to Figure 21, the vectorized HLLC, Roe, and CU solvers on the AVX machine (16 cores) gave the metrics of 5.48, 6.1, and 6.78 Mcell/s/core reaching speed-ups of 68.8×, 75.68×, and 69.6×, respectively. On the AVX2 machine (28 cores) we observed speed-ups of 96.3×,

108.4×, and 109.6× for the vectorized HLLC, Roe, and CU solvers by obtaining the metrics of 6.12, 6.98, and 8.08 Mcell/s/core, respectively. The AVX-512 machine (64 cores) shows again the significant performance increase by allowing the parallelized–vectorized HLLC, Roe, and CU solvers to achieve the metrics of 4.55, 4.57, and 5.48 Mcell/s/core or similar to speed-ups of 774.6×, 729.9×, and 742.1×, respectively. Based on this fact, a similar behavior is noticed for the single-node performance in updating the entire simulation. We take the results of the CU scheme as an example, see Figure 22. Without vectorization, the parallelized results of the AVX2 machine with 28 cores (46.80 Mcell/s) are about 1.93× significantly faster than those of the AVX-512 machine with 64 cores (24.22 Mcell/s). However, the parallelized–vectorized results of the AVX-512 machine (351 Mcell/s) now outperform the parallelized–vectorized results of the AVX2 machine (226.2 Mcell/s) by a factor of 1.55. This again shows the vectorization is highly-important for achieving better performance. For the entire simulation, our code with the vectorized solvers can achieve approximately 31–35% of the theoretical peak performance (TPP) of the AVX/AVX2 machines and 26% of the TPP of the AVX-512 machine.

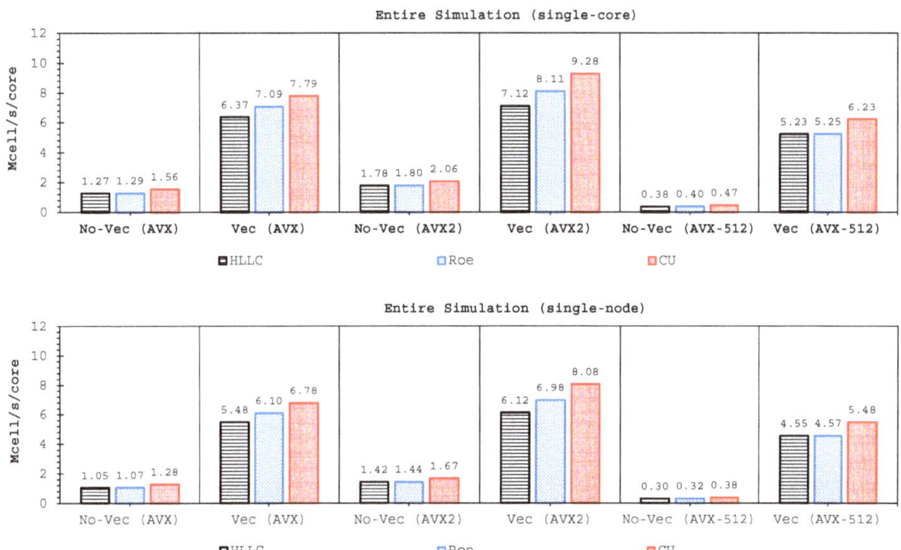

Figure 21. Comparison of performance metrics for the entire simulation.

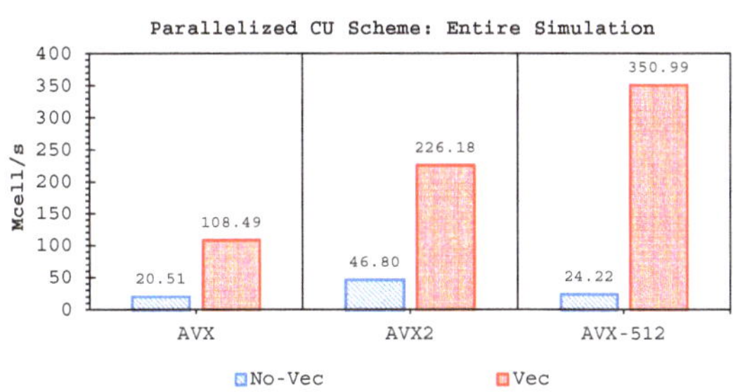

Figure 22. Single-node performance of the non-vectorized and vectorized CU scheme with parallelization (entire simulation).

We also actually studied the effect of the cell-edge reordering strategy on conserving the memory access patterns, where we compared the directive `!$omp simd simdlen(VL) aligned(var1,var2,... :Nbyte)` with the directive `!$omp simd simdlen(VL)`. Using the former, we found on all machines that the HLLC, Roe, and CU schemes averagely benefited from 1.45×, 1.5×, 1.4× more speed-ups in the edge-driven level compared to the results compiled only with the latter. Similarly, for updating the entire simulation, the HLLC, Roe, and CU solvers achieved on all machines approximately 1.41×, 1.42×, and 1.32× more speed-ups, respectively. These results reveal that the cell-edge reordering strategy proposed has helped in easing the aligned memory access pattern, thus enabling a significant performance enhancement. For the sake of brevity, these findings are not presented here.

5. Conclusions

A numerical investigation for studying the accuracy and efficiency of three common shallow water solvers (the HLLC, Roe, and CU schemes) has been presented. Four cases dealing with shock waves and wet–dry phenomenon were selected. All schemes were provided in an in-house code NUFSAW2D, the model of which was of second-order accurate in space wherever the regimes were smooth and robust when dealing with strong shock waves—and of fourth-order accurate in time. To give a fair comparison, all source terms of the 2D SWEs were treated similarly for all schemes, namely the bed-slope terms were computed separately from the convective fluxes using a Riemann-solver-free scheme—and the friction terms were computed semi-implicitly within the framework of the RKFO method.

Two important findings have been shown by our simulations. Firstly, highly-efficient vectorization could be applied to the three solvers on all hardware used. This was achieved by guided vectorization, where a cell-edge reordering strategy was employed to ease the vectorization implementations and to support the aligned memory access patterns. Regarding single-core analysis, the vectorization was shown to be able to speed-up the performance of the edge-driven level up to 4.5–6.5× on the AVX/AVX2 machines for eight data per vector and 16.7× on the AVX-512 machine for 16 data per vector—and to accelerate the entire simulation as well by up to 4–5.5× on the AVX/AVX2 machine and 13.91× on the AVX-512 machine. The superlinear speed-up in the edge-driven level especially using the AVX-512 machine could be achieved probably due to improved cache usage, thus less expensive main memory accesses. Regarding single-node analysis, our code could reach in the edge-driven level the improvements of 75.7–121.8× on the AVX/AVX2 machine while on the AVX-512 machine it achieved up to 928.9× speed-up. For updating the entire simulation, our code was able to reach speed-ups of 68.8–109.6× and 774.6× on the AVX/AVX2 and AVX-512 machines, respectively. We observed an interesting phenomenon, where without vectorization the parallelized results of the AVX2 machine outperformed those of the AVX-512 machine in both the edge-driven level and the entire simulation with a factor of up to 2×; the parallelized-vectorized results of the AVX-512 machine became, however, faster by achieving an average factor of 1.6×. This clearly shows that our reordering strategy could efficiently exploit the vectorization support of such a vector-computing machine. Supporting the aligned memory access patterns, the reordering strategy employed has helped in gaining the performances of the "only" vectorized code by averagely 1.45× and 1.4× for the edge-driven level and updating the entire simulation, respectively.

Secondly, we have shown that for the four cases simulated, strong agreements by all schemes were obtained between the numerical results and observed data, where no significant differences were shown for the accuracy. However, in the term of efficiency, the CU scheme was able to outperform the HLLC and Roe schemes with average factors of 1.4× and 1.25×, respectively. Although the vectorization was successful to significantly gain the performance of all solvers, the CU scheme still became the most efficient one among the others. According to this fact, we could conclude that the CU solver as a Riemann-solver-free scheme would in general be able to outperform the Riemann solvers (HLLC and Roe schemes) even for simulations on the next generation of modern hardware. This is because the computational procedures of the CU scheme are acceptably simple especially containing no complex branch statements (`if-then-else`) such as required by the HLLC and Roe schemes.

Since simulating shallow water flows—especially complex phenomena that require performing long real-time computations as part of disaster planning such as dam-break or tsunami cases—on modern hardware nowadays and even in the future becomes more and more common, focusing simulations only on numerical accuracy but ignoring the performance efficiency is not an option anymore. Wasting the performance is obviously undesirable due to wasting too much time for such long real-time simulations. Modern hardware offers many features for gaining efficiency, one of which is vectorization that can be regarded as the "easiest" way for benefiting from the vector-level parallelism, is thus non-trivial. However, this is not obtained for free; one should at least understand and support—due to the sophisticated memory access patterns—the vectorization concept. The cell-edge reordering strategy employed here is one of the easiest strategies to utilize the vectorization feature of modern hardware that could easily be applied to any CCFV scheme for shallow flow simulations, together with guided vectorization instead of explicitly by low-level vectorization, which might be error-prone and time-consuming. It is worth pointing out that this strategy is also applicable to any compiler with vectorization support, e.g., Gfortran. We observed that the performance obtained with Intel compiler was typically 2–3× higher than that obtained with Gfortran, which we believe is due to the correspondence of Intel compiler and Intel hardware.

We have also shown that the edge-driven level, especially the reconstruction technique and solver computations, were the most time-consuming part, which required 65–75% of the entire simulation time. This shows that some more "aggressive" optimization techniques still become a hot topic for future studies to make shallow water simulations more efficient, particularly in the edge-driven level. Finally, we conclude that this study would be useful as a consideration for modelers who are interested in developing shallow water codes.

Author Contributions: B.M.G. developed the numerical code NUFSAW2D, conceived the framework of this work, analyzed the results, and wrote the paper. R.-P.M. contributed to the practical guidance of this work and provided some corrections, especially regarding the theoretical concepts of the vectorization and parallel computing.

Funding: The first author gratefully acknowledges the DAAD (German Academic Exchange Service) for supporting his research in the scope of the Research Grants—Doctoral Programmes in Germany 2015/16 (57129429). This work was supported by the German Research Foundation (DFG) and the Technical University of Munich (TUM) in the framework of the Open Access Publishing Program.

Acknowledgments: The authors appreciate the computational and data resources provided by the Leibniz Supercomputing Centre—and are also grateful to all the anonymous reviewers, who provided many constructive comments and suggestions.

Conflicts of Interest: The authors declare no conflict of interest.

References

1. Cea, L.; Blade, E. A simple and efficient unstructured finite volume scheme for solving the shallow water equations in overland flow applications. *Water Resour. Res.* **2015**, *51*, 5464–5486. [CrossRef]
2. Hou, J.; Liang, Q.; Zhang, H.; Hinkelmann, R. An efficient unstructured MUSCL scheme for solving the 2D shallow water equations. *Environ. Model. Softw.* **2015**, *66*, 131–152. [CrossRef]
3. Duran, A. A robust and well-balanced scheme for the 2D Saint-Venant system on unstructured meshes with friction source term. *Int. J. Numer. Methods Fluids* **2015**, *78*, 89–121. [CrossRef]
4. Özgen, I.; Zhao, J.; Liang, D. Hinkelmann, R. Urban flood modeling using shallow water equations with depth-dependent anisotropic porosity. *J. Hydrol.* **2016**, *541*, 1165–1184. [CrossRef]
5. Xia, X.; Liang, Q.; Ming, X.; Hou, J. An efficient and stable hydrodynamic model with novel source term discretization schemes for overland flow and flood simulations. *Water Resour. Res.* **2017**, *53*, 3730–3759. [CrossRef]
6. Roe, P. Approximate Riemann solvers, parameter vectors and difference schemes. *J. Comput. Phys.* **1981**, *135*, 250–258. [CrossRef]
7. Toro, E. *Shock-Capturing Methods for Free-Surface Shallow Flow*; John Wiley: Chichester, UK, 2001.

8. Harten, A.; Lax, P.; van Leer, B. On upstream differencing and Godunov-type schemes for hyperbolic conservation laws. *SIAM Rev.* **1983**, *25*, 35–61. [CrossRef]
9. Davis, S. Simplified second-order Godunov-type methods. *SIAM J. Sci. Stat. Comput.* **1988**, *9*, 445–473. [CrossRef]
10. Einfeldt, B. On Godunov-type methods for gas dynamics. *SIAM J. Numer. Anal.* **1988**, *25*, 294–318. [CrossRef]
11. Toro, E.; Spruce.; M. Speares, W. Restoration of the contact surface in the HLL-Riemann solver. *Shock Waves* **1994**, *4*, 25–34. [CrossRef]
12. Kurganov, A.; Noelle, S.; Petrova, G. Semi-discrete central-upwind schemes for hyperbolic conservation laws and Hamilton-Jacobi equations. *SIAM J. Sci. Comput.* **1994**, *23*, 707–740. [CrossRef]
13. Kurganov, A.; Petrova, G. A second-order well-balanced positivity preserving central-upwind scheme for the Saint-Venant system. *Commun. Math. Sci.* **2007**, *5*, 133–160. [CrossRef]
14. Horváth, Z.; Waser, J.; Perdigão, R.; Konev, A.; Blöschl, G. A two-dimensional numerical scheme of dry/wet fronts for the Saint-Venant system of shallow water equations. *Int. J. Numer. Methods Fluids* **2015**, *77*, 159–182. [CrossRef]
15. Beljadid, A.; Mohammadian, A.; Kurganov, A. Well-balanced positivity preserving cell-vertex central-upwind scheme for shallow water flows. *Comput. Fluids* **2016**, *136*, 193–206. j.compfluid.2016.06.005. [CrossRef]
16. Delis, A.; Nikolos, I. A novel multidimensional solution reconstruction and edge-based limiting procedure for unstructured cell-centered finite volumes with application to shallow water dynamics. *Int. J. Numer. Methods Fluids* **2013**, *71*, 584–633. [CrossRef]
17. Ginting, B.; Mundani, R.P. Artificial viscosity technique: A Riemann-solver-free method for 2D urban flood modelling on complex topography. In *Advances in Hydroinformatics*; Gourbesville, P., Cunge, J., Caignaert, G., Eds.; Springer Water: Singapore, 2018; pp. 51–74, doi:10.1007/978-981-10-7218-5_4.
18. Bik, A.; Girkar, M.; Grey, P.; Tian, X. Automatic intra-register vectorization for the Intel architecture. *Int. J. Parallel Program.* **2002**, *2*, 65–98.:1014230429447. [CrossRef]
19. Nuzman, D.; Rosen, I.; Zaks, A. Auto-vectorization of interleaved data for SIMD. In Proceedings of the 27th ACM SIGPLAN Conference on Programming Language Design and Implementation, Ottawa, ON, Canada, 10–16 June 2006; pp. 132–143. [CrossRef]
20. Bader, M.; Breuer, A.; Hölzl, W.; Rettenberger, S. Vectorization of an augmented Riemann solver for the shallow water equations. In Proceedings of the 2014 International Conference on High Performance Computing & Simulation (HPCS), Bologna, Italy, 21–25 July 2014; pp. 193–201. [CrossRef]
21. Ginting, B.; Mundani, R.P. Parallel flood simulations for wet-dry problems using dynamic load balancing concept. *J. Comput. Civ. Eng. (ASCE)* **2019**, *33*, 1–18. [CrossRef]
22. Ferreira, C.; Mandli, K.; Bader, M. Vectorization of Riemann solvers for the single- and multi-layer shallow water equations. In Proceedings of the 2018 International Conference on High Performance Computing & Simulation (HPCS), Orléans, France, 16–20 July 2018; pp. 415–442. [CrossRef]
23. Liu, J.Y.; Smith, M.; Kuo, F.A.; Wu, J.S. Hybrid OpenMP/AVX acceleration of a Split HLL finite volume method for the shallow water and Euler equations. *Comput. Fluids* **2015**, *110*, 181–188. [CrossRef]
24. Ginting, B.; Mundani, R.P.; Rank, E. Parallel simulations of shallow water solvers for modelling overland flows. In Proceedings of the 13th International Conference on Hydroinformatics (HIC 2018), EPiC Series in Engineering, Palermo, Italy, 1–6 July 2018; La Loggia, G., Freni, G., Puleo, V., De Marchis, M., Eds.; Volume 3, pp. 788–799. [CrossRef]
25. Ginting, B. Central-upwind scheme for 2D turbulent shallow flows using high-resolution meshes with scalable wall functions. *Comput. Fluids* **2019**, *179*, 394–421. [CrossRef]
26. Ginting, B. A two-dimensional artificial viscosity technique for modelling discontinuity in shallow water flows. *Appl. Math. Model.* **2017**, *45*, 653–683. [CrossRef]
27. Audusse, E.; Bouchut, F.; Bristeau, M.O.; Klein, R.; Perthame, B. A fast and stable well-balanced scheme with hydrostatic reconstruction for shallow water flows. *SIAM J. Sci. Comput.* **2004**, *25*, 2050–2065. [CrossRef]
28. Gallouët, T.; Hérard, J.M.; Seguin, N. Some approximate Godunov schemes to compute shallow-water equations with topography. *Comput. Fluids* **2003**, *32*, 479–513. [CrossRef]
29. Castro, M.; Gonzales-Vida, J.; Pares, C. Numerical treatment of wet/dry fronts in shallow flows with a modified Roe scheme. *Math. Model. Methods Appl. Sci.* **2006**, *16*, 897–931. [CrossRef]
30. Yu, H.; Huang, G.; Wu, C. Efficient finite-volume model for shallow-water flows using an implicit dual time-stepping method. *J. Hydraul. Eng. (ASCE)* **2015**, *141*, 1–12. [CrossRef]

31. Soarez-Frazão.; S. Zech, Y. Experimental study of dam-break flows against an isolated obstacle. *J. Hydraul. Res.* **2007**, *45*, 27–36. [CrossRef]
32. Yu, C.; Duan, J. Two-dimensional depth-averaged finite volume model for unsteady turbulent flow. *J. Hydraul. Res.* **2012**, *50*, 599–611. [CrossRef]
33. Briggs, M.; Synolakis, C.; Harkins, G.; Green, D. Laboratory experiments of tsunami runup on a circular island. *Pure Appl. Geophys.* **1995**, *144*, 569–593. [CrossRef]
34. Nikolos, I.; Delis, A. An unstructured node-centered finite volume scheme for shallow water flows with wet/dry fronts over complex topography. *Comput. Methods Appl. Mech. Eng.* **2009**, *198*, 3723–3750. [CrossRef]
35. Available online: https://coastal.usc.edu/currents_workshop/index.html (accessed on 25 September 2018).
36. Arcos, M.; LeVeque, R. Validating velocities in the GeoClaw tsunami model using observations near Hawaii from the 2011 Tohoku tsunami. *Pure Appl. Geophys.* **2014**, *17*, 849–867. [CrossRef]
37. Available online: https://sandstorm.cie.bgu.tum.de/wiki/index.php/Main_Page (accessed on 25 September 2018).
38. Available online: https://www.lrz.de (accessed on 25 September 2018).

Article

2D Numerical Modeling on the Transformation Mechanism of the Braided Channel

Shengfa Yang and Yi Xiao *

National Inland Waterway Regulation Engineering Research Center, Chongqing Jiaotong University, Chongqing 400074, China; ysf777@163.com

* Correspondence: xymttlove@163.com; Tel.: +86-023-6265-4621

Received: 25 July 2019; Accepted: 24 September 2019; Published: 28 September 2019

Abstract: This paper investigates the transformation mechanism between different channel patterns. A developed 2D depth-averaged numerical model is improved to take into account a bank vegetation stress term in the momentum conservation equation of flow. Then, the extended 2D model is applied to duplicate the evolution of channel pattern with variations in flow discharge, sediment supply and bank vegetation. Complex interaction among the flow discharge, sediment supply and bank vegetation leads to a transition from the braided pattern to the meandering one. Analysis of the simulation process indicates that (1) a decrease in the flow discharge and sediment supply can lead to the transition and (2) the riparian vegetation helps stabilize the cut bank and bar surface, but is not a key in the transition. The results are in agreement with the criterion proposed in the previous research, confirming the 2D numerical model's potential in predicting the transition between different channel patterns and improving understanding of the fluvial process.

Keywords: fluvial process; bank vegetation; channel pattern; 2D numerical model

---

1. Introduction

Channel pattern refers to the limited reaches of the river that can be defined as straight, meandering or braided. While long, straight rivers seldom occur in nature; meandering and braided rivers are common [1]. The transformation of channel patterns take place in response to variations in different variables, which can be grouped into four categories: (i) Dynamic flow, (ii) shape and characteristics of the channel, (iii) sediment load and (iv) bed and bank material [2]. A sound understanding of the relationship between the control variables and channel pattern is fundamental to the development of improved management strategies in braided rivers [3]. The laboratory flume experiments have shed much light on the dynamic behavior of a wide braided river to a single-thread channel [4–10]. Various criteria have been proposed on the response of channel morphology to control variables [11–14]. Quantitative inconsistencies in both the coefficients and exponents of discriminant functions have resulted from the use of different measures of slope and discharge, as well as differences in the definitions of the transition between channel patterns [15–17].

With the rapid developments of numerical and mathematics methods in fluid mechanics, multiple-mathematics models have become important tools for investigating dynamic interactions in evolving braid units. The development of physically based theories, which attempt to relate pattern and process in a predictive manner, offer improved insight into the primary variables controlling channel pattern. Models based on linearized physics-based equations [18–21] and 2D nonlinear physics-based morphological models [22–25] have been established to simulate the braided channel evolution. Cellular models [26–29], 2D and 3D flow-sediment numerical models [30–35] have been developed to model braided rivers. Although various computational studies on the formation of braided rivers are available, few preliminary numerical studies of the transformation process from the braided to meandering pattern are offered [36], to discuss the interactions of multiple factors, such as flow conditions, sediment characteristics and bank stability.

The primary objective of this study is to investigate the dynamic process of the transformation between different channel patterns with different control variables. The original 2D numerical model takes the vegetation term into the flow momentum equation, and is verified in the middle section of the Yangtze River. Subsequently, a conceptual braided channel is established in the numerical experiment, control factors as flow discharge, sediment supply and bank vegetation are considered in the simulation of the transition from the braided to the meandering channel. The proposed criteria were applied to discuss the transition process between the braided and meandering channel, the results agree well with the previous research. It demonstrates that the 2D numerical model can be applied to improve understanding of patterning processes under different scenarios.

2. Numerical Model

2.1. Model Description

The 2D numerical model incorporates the hydrodynamic, sediment transport and river morphological adjustment sub-model. It is solved in the orthogonal curvilinear grid system by using the Beam and Warming alternating-direction implicit (ADI) scheme. The sediment transport submodel includes the influence of non-uniform sediment with bed surface armoring and a correction for the direction of bed-load transport due to secondary flow and transverse bed slope. The bank erosion submodel incorporates a simple simulation method for updating bank geometry during either degradational or aggradational bed evolution. The details of the developed 2D model can be found in Xiao et al. [37], and verified in the physical meandering channel and the upstream of the Yangtze River [38].

2.2. Consideration of the Riparian Vegetation Influence

The significance of riparian vegetation as a control of river form and process is increasingly being recognized in fluvial research. In this study, the hydrodynamic portion of the 2D numerical model was upgraded to incorporate the effects of riparian vegetation.

The equilibrium equation for the riparian vegetation zones herein can be introduced by Ikeda and Izumi [39] in the form:

$$\frac{\tau}{\cos\theta} = \rho g H S - D_r + \frac{d}{dy}\int_0^H \left(-\rho\overline{u'v'}\right)dz, \tag{1}$$

where τ is the total shear stress near the river bank (Pa); D_r is the vegetation stress term (Pa); v', u' are the fluctuating velocity in the longitudinal and transverse direction (m/s), respectively; S is the slope, H is the averaged water depth (m) and θ is the inclination of the location, often $\theta \approx 0$, Equation (1) can be reduced to:

$$\begin{gathered} \tau = \rho g H S - D_r + \tfrac{d}{dy}\int_0^H \left(-\rho\overline{u'v'}\right)dz \\ D_r = \tfrac{1}{2}\rho C_D \overline{u}^2 \tfrac{aH}{\cos\theta} \\ \tau = \tau_j^L + \tau^T - D_r. \end{gathered} \tag{2}$$

Let $p^v = D_r$, substitute it to the momentum conservation equation of flow in the Cartesian coordinate system as:

$$\frac{\partial}{\partial t}(\rho u_i) + \frac{\partial}{\partial x_j}(\rho u_i u_j) = \rho f_i - \frac{\partial p}{\partial x_i} + \frac{\partial \tau_{ij}}{\partial x_j} - \frac{\partial p^v}{\partial x_i}, \tag{3}$$

$$\frac{\partial p^v}{\partial x_i} = \frac{\partial\left(\frac{1}{2}\rho C_D \overline{u}^2 \frac{aH}{\cos\theta}\right)}{\partial x_i} = \frac{1}{2}\rho C_D \frac{aH}{\cos\theta}\overline{u}u_i \; i = 1,2, \tag{4}$$

$$\overline{u} = \sqrt{\sum_i u_i^2} \quad i = 1,2 \tag{5}$$

where p^v should satisfy the additional condition in all directions as: $p^v = \sum_{i=1}^{2}\left(\frac{\partial p^v}{\partial x_i}\right)^2$; $\overline{u}$ is the depth-averaged flow velocity (m/s); u_i is the flow velocity in the i-direction (m/s); a is the vegetation density (m^{-1}), defined as $a = d/\left(l_x l_y\right)$, d is the radius of the vegetation (m) and l_x and l_y are the distance of vegetation in the longitudinal and transverse directions (m).

C_D is the drag coefficient of vegetation. Consider the influence range of the vegetation coefficient, let C_D = 1.5 when the vegetation zones near the river bank [39]; if the zones of vegetation are in the river channel, we assumed the influence of vegetation was proportionate to the distance from the channel center in the form:

$$\begin{array}{ll} C_D = 0 & x = l \\ C_D = 1.5 - 1.5x/l & 0 < x < l \\ C_D = 1.5 & x = 0 \end{array} \tag{6}$$

where l is the distance from the river bank to the channel center (m); x is the distance from the computed point to the river bank (m).

In this study, we substituted Equations (4) and (5) to the 2D depth-averaged momentum conservation equation of flow in the orthogonal curvilinear coordinate system as follows:

$$\begin{aligned} &\frac{\partial q}{\partial t} + \beta\left(\frac{1}{J}\frac{\partial (h_2 qU)}{\partial \xi} + \frac{1}{J}\frac{\partial (h_1 pU)}{\partial \eta} - \frac{pV}{J}\frac{\partial h_2}{\partial \xi} + \frac{qV}{J}\frac{\partial h_1}{\partial \eta}\right) - fp + \frac{gH}{h_1}\frac{\partial Z}{\partial \xi} + \frac{qg|\overline{q}|}{(CH)^2} = \frac{v_e H}{h_1}\frac{\partial E}{\partial \xi} - \frac{v_e H}{h_2}\frac{\partial F}{\partial \eta} + \frac{1}{J}\frac{\partial (h_2 D_{11})}{\partial \xi} \\ &\quad + \frac{1}{J}\frac{\partial (h_1 D_{12})}{\partial \eta} + \frac{1}{J}\frac{\partial h_1}{\partial \eta}D_{12} - \frac{1}{J}\frac{\partial h_2}{\partial \xi}D_{22} - \frac{1}{2}\rho C_D \frac{aH}{\cos\theta}\sqrt{U^2 + V^2}\frac{y_\eta h_1 U - y_\xi h_2 V}{J} \\ &\frac{\partial p}{\partial t} + \beta\left(\frac{1}{J}\frac{\partial (h_2 qV)}{\partial \xi} + \frac{1}{J}\frac{\partial (h_1 pV)}{\partial \eta} + \frac{pU}{J}\frac{\partial h_2}{\partial \xi} - \frac{qU}{J}\frac{\partial h_1}{\partial \eta}\right) + fq + \frac{gH}{h_2}\frac{\partial Z}{\partial \eta} + \frac{pg|\overline{q}|}{(CH)^2} = \frac{v_e H}{h_2}\frac{\partial E}{\partial \eta} + \frac{v_e H}{h_1}\frac{\partial F}{\partial \xi} + \frac{1}{J}\frac{\partial (h_2 D_{12})}{\partial \xi} \\ &\quad + \frac{1}{J}\frac{\partial (h_1 D_{22})}{\partial \eta} - \frac{1}{J}\frac{\partial h_1}{\partial \eta}D_{11} + \frac{1}{J}\frac{\partial h_2}{\partial \xi}D_{12} - \frac{1}{2}\rho C_D \frac{aH}{\cos\theta}\sqrt{U^2 + V^2}\frac{x_\xi h_2 V - x_\eta h_1 U}{J} \end{aligned} \tag{7}$$

where h_1 and h_2 are the lame coefficients in the ξ and η direction, respectively; U and V are the depth-averaged flow velocity components in the ξ and η direction; the unit discharge vector is $\overline{q} = (q, p) = (UH, VH)$; z is the water level relative to the reference plane; β is the correction factor for non-uniformity of the vertical velocity profile; f is the Coriolis parameter, which was neglected in this study; g is the gravitational acceleration; C is the Chezy coefficient; v_e is the depth mean effective vortex viscosity, z_s and z_b are the dependent water levels at the water surface and channel bed, respectively.

2.3. Verification

The extended 2D numerical model was applied to a 102 km long, 'S' shaped channel section in the middle Yangtze River, and the bank along the river from Shashi to Shishou is protected by the riparian vegetation. An orthogonal curvilinear coordinate system was applied with a total of 600 × 115 grids in the computational domain and a time interval of t = 8 s (Figure 1). The angles between the ξ and η grid lines were 88° and −92°, except for some grids close to the banks. The grid spacing was 100–180 m in the ξ direction and 35–45 m in the η direction. Observed daily water discharge and sediment load at the inlet were used as boundary conditions and bed contour maps dated September 2002 was the initial topography [40]. Calculation of suspended load was divided to eight group ranging from 0.005 to 1 mm in diameter (Table 1). The sediment gradation in bed materials (Table 2), transport capacity for various size groups, and river topography were adjusted every 24 h. The thickness of active layers were L_a = 15 m. A real time period of two years was simulated, and the calculated results of flow velocity, water stage and morphological changes were compared with the measured data.

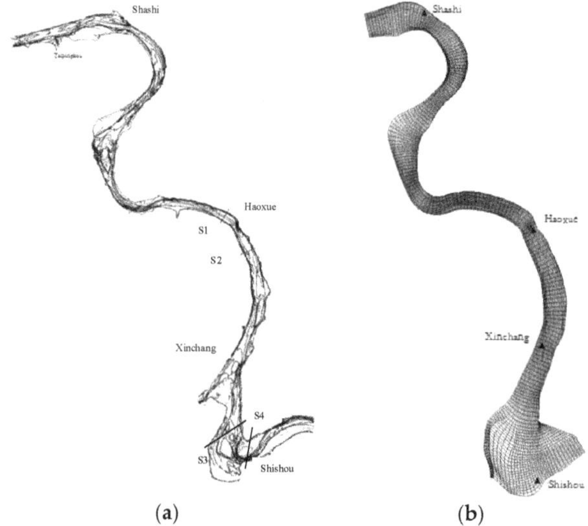

Figure 1. Layout of the field study reach section and its computational mesh. (**a**) layout of the study river section; (**b**) computational mesh.

Table 1. The fraction of suspended load being simulated.

No.	1	2	3	4	5	6	7	8
Size (mm)	0.004	0.008	0.016	0.031	0.062	0.125	0.25	0.5
Proportion	30	12.7	13.4	14.6	13.1	8.2	6.5	1.5

Table 2. The fraction of bed material.

No.	Group Percentage of Bed Materials									D_{50} (mm)	Year
	0.004	0.008	0.016	0.03	0.062	0.125	0.25	0.5	1		
%	0	0	0	0.1	1.1	13.2	55.3	30	0.3	0.193	2002

Comparison of observed and calculated cross-sectional profile of depth averaged stream-wise velocity for various discharges in November 2003 is shown in Figure 2, calculated depth-averaged velocities were consistent with the observed asymmetrical velocity patterns, and the relative error near the bank vegetation area was below 6%. Figure 3 shows the comparison of the measured and calculated water stages at two hydrometric stations during September 2002–July 2004, which indicate good agreements between simulations and measurements.

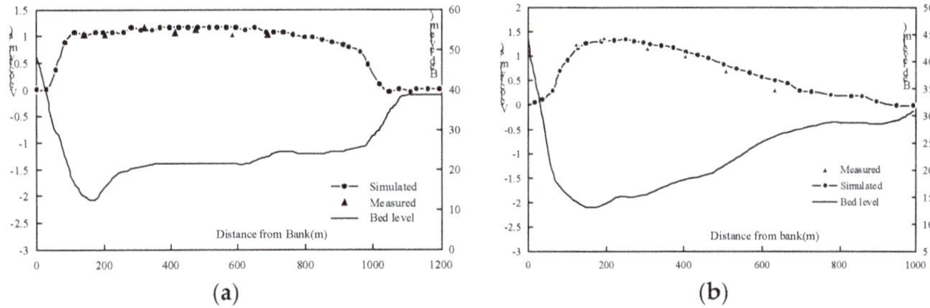

Figure 2. Measured and calculated cross-sectional profiles of depth-averaged velocity. (**a**) cross section S1; (**b**) cross section S2.

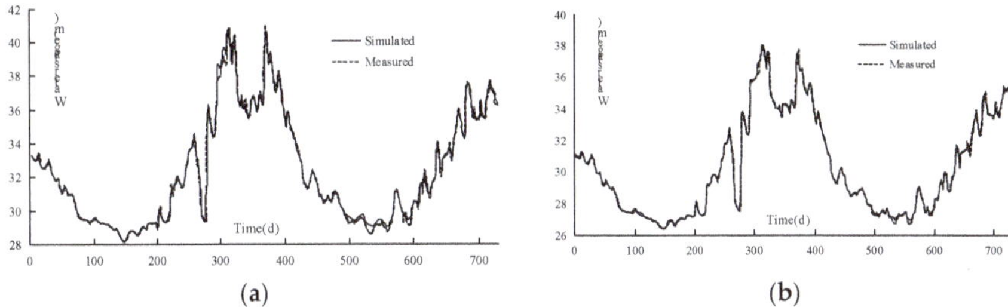

Figure 3. Comparison of the water stages at two control stations. (**a**) Shashi station; (**b**) Xinchang station.

Table 3 lists the measured and calculated total amount of deposition or scour. It indicates that the largest discrepancy between observed and calculated of results was found in the entrance section from Taipingkou-shashi, possibly due to the uncertainties introduced by the initial and boundary conditions. Figure 4 is a comparison between the calculated and measured scour and deposition depths. It can be seen that except the entrance section, the predicted pattern of scour and deposition agreed well with observations if reliable information of bank material, riparian vegetation and bed material size could be obtained. A comparison of changes of the bed level at the typical cross sections shows that as time progressed, the pattern of the cross sections tended to the measurements with acceptable ranges of error (Figure 5).

Table 3. Measured and calculated volumes of deposition (+) or scour (−).

River Section	Total Distance (km)	Section Length (km)	Measured (10^6 m^3)	Calculated (10^6 m^3)
Taipingkou-Shashi	8.47	8.47	−827.26	−1185.91
Shashi-Haoxue	58.65	50.19	−1705.39	−1730.82
Haoxue-Xinchang	73.62	14.96	−1353.62	−924.21
Xinchang-Shishou	93.38	19.76	−1508.87	−1719.86

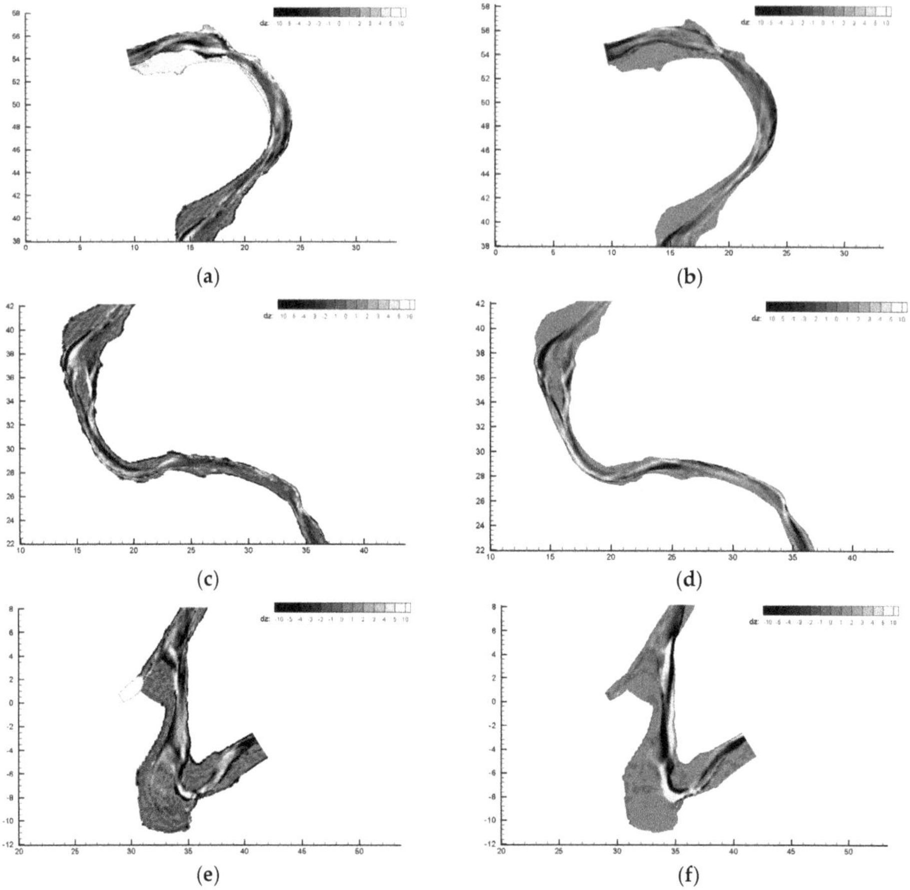

(**a**) (**b**)

(**c**) (**d**)

(**e**) (**f**)

Figure 4. Calculated and measured scour or deposition depths of reach section (x: the distance from the x-direction/m; y: the distance from the y-direction/m; dz: the bed level changes/m). (**a**) Measured; (**b**) calculated; (**c**) measured; (**d**) calculated; (**e**) measured and (**f**) calculated.

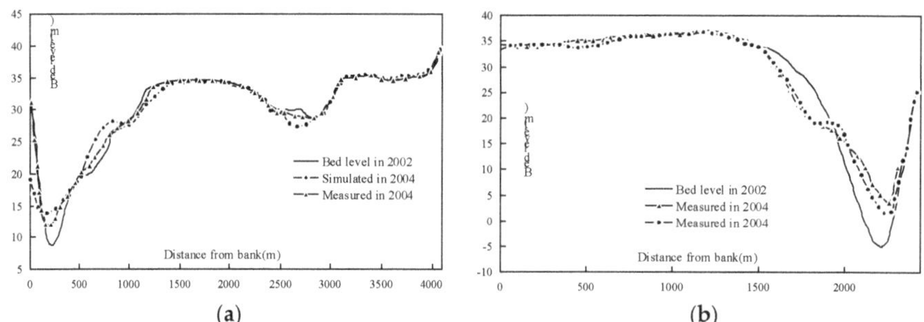

(**a**) (**b**)

Figure 5. Measured and calculated bed deformation at various typical cross sections. (**a**) cross section S3 and (**b**) cross section S4.

3. Numerical Modeling on the Transformation of Braided and Meandering Channel

3.1. Formation of the Braided Channel

The conceptual channel was 10,000 m long and 300 m wide, and the grid system of 400 × 80 nodes was generated. The initial bed was flat with a 0.4% slope, the medium grain size of the sediment supply and the bed material was 0.1 mm. The inlet water discharge and sediment feed rate are provided in Table 4, and the outlet water level was constant during the simulation, the repose of the sediment $\varphi' = 14$, and the lateral erosion coefficient of the bank as C = 0.011. The computational time interval $\Delta t = 6$ s, and the simulation time period was 720 days.

Table 4. The experimental conditions.

Time Period	Time (d)	Discharge (m^3/s)	The Medium Grain Size (mm)	Sediment Supply (kg/m^3)
1	360	150	0.1	1
2	360	300	0.1	5

Figure 6 depicts an unstable braided river pattern after 720 days. Two control factors contributed to the formation of the braided channel: Large and sudden variation in discharge resulted in broadened channel cross-sections; large sediment supply led to aggradation up and down in the upper section of the stream and the initially symmetric inflow became almost asymmetrical and formed point bars or migrating central bars. It illustrated that a fluctuation in the controls would induce changes of the braided channel pattern to another pattern.

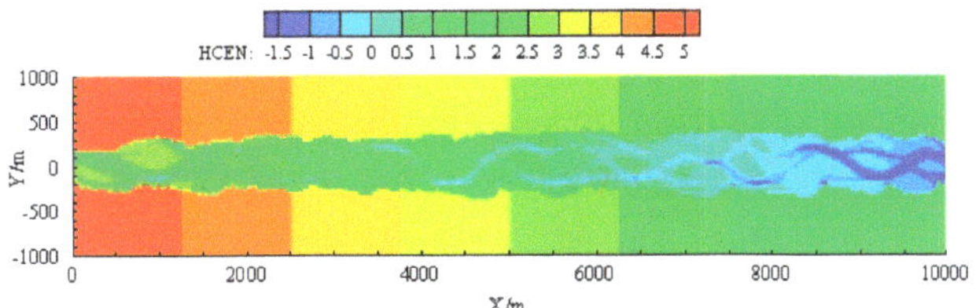

Figure 6. Layout of the conceptual channel after 720 days.

3.2. The Transformation of the Braided Channel under Control Variables

Based on the simulated braided river, four numerical experiments were performed including the effect of water discharge, sediment supply and bank vegetation. The experimental conditions can be seen in Table 5.

Table 5. The experimental conditions.

No.	Flow Discharge (m^3/s)	Sediment Supply (kg/m^3)	Bank Vegetation	Time (d)
1	150	5	Yes	600
2	300	1	No	600
3	300	5	No	600
4	150	1	Yes	600

Figure 7 depicts the final planform of the braided river for runs No. 1, 2 and 3. In run No. 1, reduction of the discharge led to a weak sediment transport capacity, sedimentation took place in the branch channel and a new main channel was formed in the upper section. With time processes, aggradation resulted in higher bed elevations above the initial bed profile in the upstream, led to an increase of the stream power in the downstream and a broad, island braided channel was formed

(Figure 7a). The braided channel in run No. 2 also transferred to a meandering channel in the upstream with different mechanisms compared with run No. 1: A reduction of sediment load resulted in less aggradation and bed scour in the upper part, and might be a key factor in the formation of a straight channel pattern with no island-bars in the downstream (Figure 7b). Figure 7c shows that bank vegetation enhanced the strength of banks, stabilized the channel, held on the sediment and the plan view seemed like that of run No. 2. As shown in Figure 7d, the planform of run No. 4 was obtained by the contribution of the influence of discharge, sediment supply and bank vegetation. It can be seen that the channel transformed to a single thread channel pattern differing from the other three numerical experiments, especially in the downstream; the reach downstream was sketched, where the wetted and active branches were marked off.

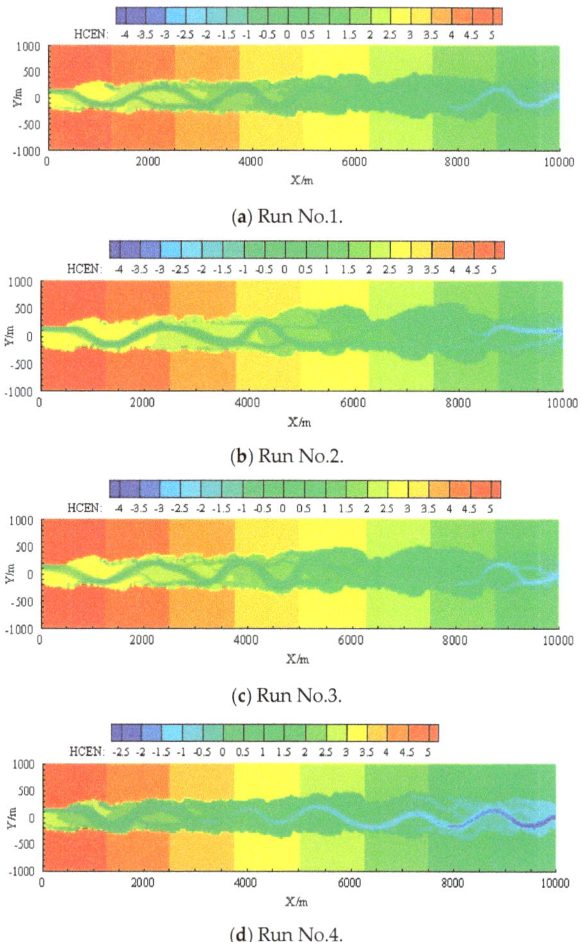

Figure 7. Layout of the experimental channel after 600 days (HCEN: Bed level/m). (**a**) Run No. 1; (**b**) Run No. 2; (**c**) Run No. 3; (**d**) Run No. 4.

4. Discussion

4.1. The Cross Section Change

Figure 8 shows the comparison of the bed deformations between runs No. 1–3 and the initial braided river at the 6000 m cross section. As decreasing the discharge and sediment load respectively in run No.1 and 2, the main channel shifted to the right bank as the sand bars growing at the left bank; the shape of the cross section transit from "W" to "U"; the width ratio was lower and the depth of the channel in run No. 3 was deeper than that of run No. 1–2, it illustrated that the vegetation could increase tensile and shear strength, gave adequate time and conditions for development, such stabilization allows the existence of relatively steep cut banks, and might hinder the lateral migration of channels [41].

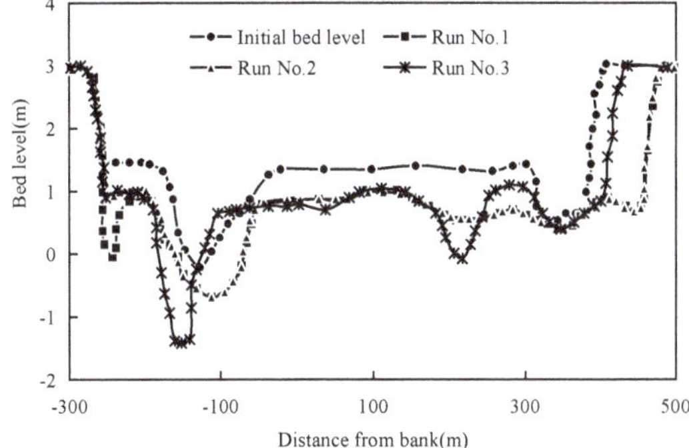

Figure 8. Comparison of bed deformation at the 6000 m cross-section.

4.2. The Channel Planform Change

The quantified parameters characterizing run No. 1–4 were obtained in Table 3. "Braided -channel ratio" B was used to describe the development of multiple channels from a channel belt as follows [42]:

$$B = L_{ctot}/L_{cmax}, \tag{8}$$

where L_{ctot} is the sum of the mid-channel lengths of all the segments of primary channels in a reach and L_{cmax} is the mid-channel length of the same channel.

Table 6 shows the braiding and meandering parameters for run No. 1–4. Due to the similar plan view in run No. 2 and run No. 3, one could see the values of the sinuosity (P) and braided-channel ratio (B) tended to correlate negatively with the reduction of breaches. Figure 9 presents the sketch of the braided reach for the initial and run 1, 2 and 4. Theoretically, if a reach has only a single channel, with no braids, the braided-channel ratio (B) would approach 1 as the sinuosity (P) of the river section has the minimum value of unity.

Table 6. The parameters of the braided reach.

No.	Number of Breaches	Braided-Channel Ratio (B)	Sinuosity (P)
Run No. 1	6	2.11	1.06
Run No. 2	5	1.9	1.00
Run No. 3	4	1.97	1.01
Run No. 4	2	1.22	1.35

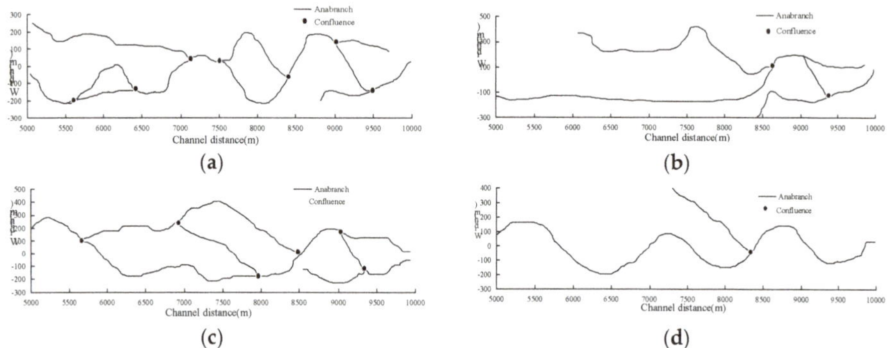

Figure 9. Sketch of the braided reach for initial and run 1–4. (**a**) Initial reach; (**b**) run 1; (**c**) run 2 and (**d**) run 4.

A large portion of branches exhibited morphological activity, with seven branches in initial reach as shown in Figure 9a, the number of branches was reduced to two in run No. 4 while the channel pattern became the meandering (Figure 9d). The results reflected that the value of P would decrease with the channel belts intersect each other, and the channel belts developing along the single-channel, meandering arm had higher sinuosity. The flow field of run No. 4 was plotted in Figure 10, including the velocity and bed elevation; it can be seen that reduction of the inlet discharge and sediment supply led to a meandering flow path. The results demonstrate that the discharge and sediment supply played a significant role in the transformation mechanism of channel patterns, which agreed qualitatively with the previous work on this topic [10].

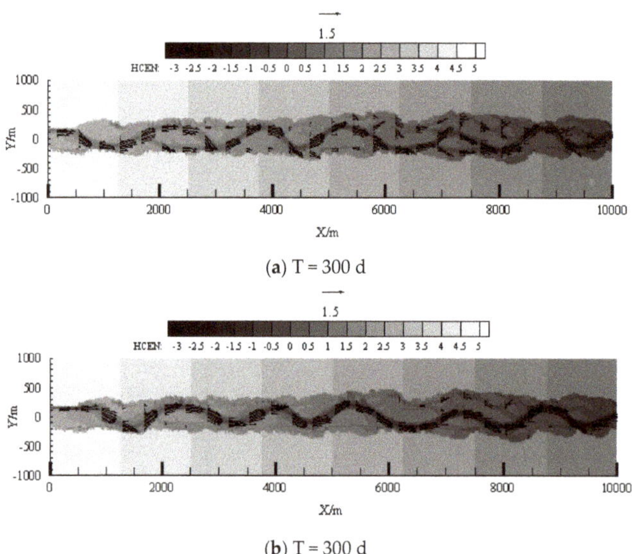

Figure 10. Temporal changes in flow field including velocity and bed elevation of Run No. 4. (**a**) T = 300 d and (**b**) T = 300 d.

4.3. Comparison with the Empirical Dimensionless Braiding Criterion

Just over 50 years ago Leopold and Wolman [4] published their classic analysis of alluvial river patterns. The number of channel classification schemes increased rapidly in the following decades. The single most cited component of Leopold and Wolman is the empirical expression for the meandering-braiding threshold slope, S^*:

$$S^* = 0.0125 \times Q^{-0.44}, \tag{9}$$

where Q is the bankfull discharge (m^3/s). Channel pattern is determined at least in part by both the rate and mode of sediment transport, an obvious shortcoming of Equation (9) is the absence of bed material size. Henderson [43] reanalyzed the Leopold and Wolman data and derived an equivalent expression:

$$S^* = 0.52 \cdot D_{50}^{1.14} \times Q^{0.44}, \tag{10}$$

where D_{50} is the median bed surface grain size (m). Equation (10) can be expressed using the dimensionless discharge defined by Parker [44]. The dimensionless discharge, Q^*, is given by:

$$Q^* = \frac{Q}{D_{50}^2 \sqrt{(s-1)gD_{50}}}, \tag{11}$$

where s is the specific gravity of the sediment grains. Millar [45] found that for channels where the relative bank strength does not change appreciably with the channel size, and then combined regime theory with a linear stability model to generate a morphodynamic power functions that describe the threshold slope as a function of Q:

$$S^* = 0.00957\mu' Q^{*-0.25}, \tag{12}$$

where μ' is the dimensionless relative bank strength given by the ratio of the critical shear stress for entrainment of the channel banks to the critical shear stress for the channel bed.

Figure 11 shows the temporal changes of the braiding criterion under four different simulation conditions. It can been seen that the data of run No. 1–3 were located in the upper bound for the braided channels, and run No. 4 data was in the lower bound for braided stream. It indicates that the relative bank strength strongly influenced channel geometry, and so for channels where the banks were more resistant than the bed, because of vegetation, we could expect a single-thread channel to persist in a region where braiding would otherwise be expected to occur [46].

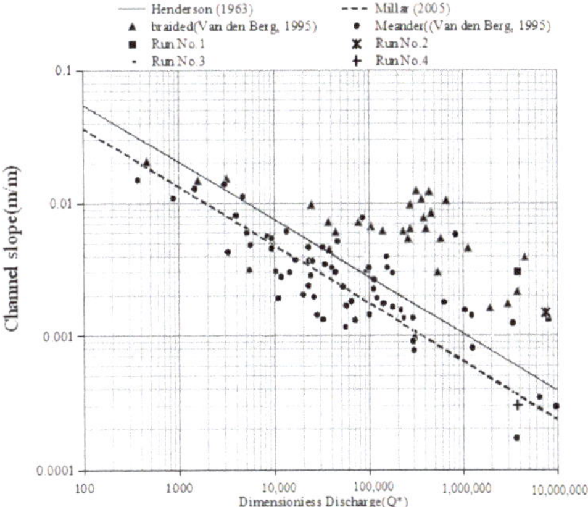

Figure 11. Dimensionless braid thresholds with the numerical experiment data.

5. Conclusions

This paper presented research on the transformation mechanisms from a braided to meandering pattern by a numerical approach. A 2D depth-averaged hydrodynamic model for hydrodynamic, sediment transport and river morphological adjustment was applied in the numerical experiment. A conceptual braided channel and its transformation with different control factors were simulated to study the mechanics of fluvial process. It demonstrated that the tendency of the research on the mechanisms of fluvial processes might be regarded as a combination of the theoretical study with numerical models in future. Further studies are needed to research the fundamental equation that governs the evolution of alluvial river, which has not been fully understood to ensure the availability of the numerical model.

Author Contributions: Numerical model and experiment, S.Y. and Y.X.; analysis and manuscript preparation, S.Y. and Y.X.

Funding: This research was founded by [the National Natural Science Foundation of China] grant number [51679020].

Acknowledgments: We greatly appreciate anonymous reviewer's constructive comments which helped to improve the quality of our manuscript.

Conflicts of Interest: The authors declare no conflict of interest.

Abbreviations

τ_{ij}	The shear-stress tensor
τ	The total shear stress near the river bank
S	The slope of the water surface
u, v	The time-averaged flow velocity components in the Cartesian coordinate system
a	The vegetation density
l_x, l_y	The distance of vegetation in the longitudinal and transverse direction
ξ, η	The orthogonal curvilinear coordinates
h_1, h_2	The Lamé coefficients
J	The Jacobian of the transformation $J = h_1 h_2$
Z	The water level relative to the reference plane
H	The averaged water depth
U, V	The depth-averaged velocity components in the ξ and η directions
β	The correction factor for the non-uniformity of the vertical velocity
f	The Coriolis parameter
g	The gravitational acceleration
C	The Chezy coefficient
v_e	The depth mean effective vortex viscosity
$D_{11}, D_{12}, D_{21}, D_{22}$	The depth-averaged dispersion stress terms
z_s, z_b	The dependent water levels for the water surface and channel bed
θ	The inclination of the location
D_r	The vegetation stress term
k	von Karman constant
Δt	The time increment
B	The "braided-channel ratio"
L_{ctot}	The sum of the mid-channel lengths of all the segments of primary channels in a reach
L_{cmax}	The mid-channel length of the same channel
S^*	The meandering-braiding threshold slope
Q	The bankfull discharge
Q^*	The dimensionless discharge
D_{50}	The median grain size

References

1. Richardson, W.R.; Thorne, C.R. Multiple thread flow and channel bifurcation in a braided river: Brahmaputra-jamuna river, Bangladesh. *Geomorphology* **2001**, *38*, 185–196. [CrossRef]
2. Biedenharn, D.S.; Watson, C.C.; Thorne, C.R. Fundaments of fluvial geomorphology. In *Sediment Engineering: Processes, Measurements. Modelling and Practice*; Garcia, M.H., Ed.; ASCE: New York, NY, USA, 2008; pp. 355–386.
3. Nicholas, A.P. Modelling bedload yield in braided gravel bed rivers. *Geomorphology* **2000**, *36*, 89–106. [CrossRef]
4. Leopold Luna, B.; Wolman, M. *Gordon. River Channel Patterns: Braided, Meandering and Straight*; U.S. Government Printing Office: Washington, DC, USA, 1957.
5. Acker, P.T.; Charlton, F.G. The geometry of small meandering streams. *Proc. Inst. Civil Eng.* **1971**, *172*, 289–317.
6. Schumm, S.A.; Khan, H.R. Experimental study of channel patterns. *Geol. Soc. Am. Bull.* **1972**, *83*, 1755–1770. [CrossRef]
7. Ikeda, H. A study of the formation of sand bars in an experimental flume. *Geogr. Rev. Jpn.* **1973**, *46*, 435–452. [CrossRef]
8. Ikeda, H. On the bed configuration in alluvial channels; their types and condition of formation with reference to bars. *Geogr. Rev. Jpn.* **1975**, *48*, 712–730. [CrossRef]
9. Ashmore, P.E. Laboratory modeling of gravel braided stream morphology. *Earth Surf. Process. Landf.* **1982**, *7*, 201–225. [CrossRef]
10. Ashmore, P.E. How do gravel-bed rivers braid? *Can. J. Earth Sci.* **1991**, *28*, 326–341. [CrossRef]
11. Gill, M.K. Erosion of sand beds around spur dikes. *J. Hydraul. Div.* **1972**, *98*, 1587–1602.
12. Klingeman, P.C.; Kehe, S.M.; Owusu, Y.A. *Steambank Erosion Protection and Channel Scour Manipulation Using Rockfill Dikes and Gabions*; Technical Report; Water Resources Research Institute: Corvallis, OR, USA, 1984.
13. Kuhnle, R.A.; Alonso, C.; Shields, F.D. Geometry of scour holes associated with 90 spur dikes. *J. Hydraul. Eng.* **1999**, *125*, 972–978. [CrossRef]
14. Eaton, B.C.; Millar, R.G.; Davidson, S. Channel patterns: Braided, anabranching, and single-thread. *Geomorphology* **2010**, *120*, 353–364. [CrossRef]
15. Van den Berg, J.H. Prediction of alluvial channel pattern of perennial rivers. *Geomorphology* **1995**, *12*, 259–279. [CrossRef]
16. Alabyan, A.M.; Chalov, R.S. Types of river channel patterns and their natural controls. *Earth Surf. Process. Landf.* **1998**, *23*, 467–474. [CrossRef]
17. Beechie, T.J.; Liermann, M.; Pollock, M.M.; Baker, S.; Davies, J. Channel pattern and river-floodplain dynamics in forested mountain river systems. *Geomorphology* **2006**, *78*, 124–141. [CrossRef]
18. Ikeda, S.; Parker, G.; Sawai, K. Bend theory of river meanders, 1, Linear development. *J. Fluid Mech.* **1981**, *112*, 363–377. [CrossRef]
19. Johannesson, H.; Parker, G. Linear theory of river meanders. *Water Resour. Monogr.* **1989**, *12*, 181–213.
20. Zolezzi, G.; Seminara, G. Downstream and upstream influence in river meandering. Part1: General theory and application to overdeepening. *J. Fluid Mech.* **2001**, *438*, 183–211. [CrossRef]
21. Crosato, A. Analysis and Modelling of River Meandering. Ph.D. Thesis, Delft University of Technology, Delft, The Netherlands, 2008.
22. Osman, A.M.; Thorne, C.R. Riverbank stability analysis, I: Theory. *J. Hydraul. Eng.* **1988**, *114*, 134–150. [CrossRef]
23. Mosselman, E. Morphological modeling of rivers with erodible banks. *Hydrol. Process.* **1998**, *12*, 1357–1370. [CrossRef]
24. Darby, S.E.; Alabyan, A.M.; Van de Wiel, M.J. Numerical simulation of bank erosion and channel migration in meandering rivers. *Water Resour. Res.* **2002**, *38*, 1–21. [CrossRef]
25. Duan, J.G.; Julien, P.Y. Numerical simulation of meandering evolution. *J. Hydrol.* **2010**, *391*, 34–46. [CrossRef]
26. Murray, A.B.; Paola, C. A cellular model of braided rivers. *Nature* **1994**, *371*, 54–57. [CrossRef]
27. Murray, A.B.; Paola, C. Modelling the effect of vegetation on channel pattern in bedload rivers. *Earth Surf. Proc. Land* **2003**, *2*, 131–143. [CrossRef]

28. Paola, C. Modelling stream braiding over a range of scales. In *Gravel Bed Rivers*; Mosley, M.P., Ed.; New Zealand Hydrological Society: Wellington, New Zealand, 2001; pp. 111–146.
29. Thormas, R.; Nicholas, A.P. Simulation of braided river flow using a new cellular routing scheme. *Geomorphology* **2002**, *43*, 179–196. [CrossRef]
30. Takebayashi, H.; Okabe, T. Numerical modeling of braided streams in unsteady flow. *Water Manag.* **2009**, *162*, 189–198.
31. Bridge, J.S.; Lunt, I.A. Depositional models of braided rivers. In *Braided Rivers: Process, Deposits, Ecology and Management*; Wiley: Hoboken, NJ, USA, 2009.
32. Jang, C.L.; Shimizu, Y. Numerical analysis of braided rivers and alluvial fan deltas. *Eng. Appl. Comput. Fluid Mech.* **2009**, *1*, 390–395. [CrossRef]
33. Schuurman, F.; Kleinhans, M. Self-formed braided bar pattern in a numerical model. In *River, Coastal and Estuarine Morphodynamics*; Springer: Berlin/Heidelberg, Germany, 2011.
34. Lotsari, E.; Wainwright, D.; Corner, G.D.; Alho, P.; Kayhko, J. Surveyed and modeled one-year morphodynamics in the braided lower Tana River. *Hydrol. Process.* **2014**, *28*, 2685–2716. [CrossRef]
35. Karmaker, T.; Dutta, S. Prediction of short-term morphological change in large braided river using 2D numerical model. *J. Hydraul. Eng.* **2016**, *142*, 04016039. [CrossRef]
36. Crosato, A.; Mosselman, E. Simple physics-based predictor for the number of river bars and the transition between meandering and braiding. *Water Resour. Res.* **2009**, *44*, W03424. [CrossRef]
37. Xiao, Y.; Shao, X.J.; Wang, H.; Zhou, H. Formation process of meandering channel by a 2D numerical simulation. *Int. J. Sediment. Res.* **2012**, *3*, 306–322. [CrossRef]
38. Xiao, Y.; Yang, S.F.; Su, L. Fluvial sedimentation of the permanent backwater zone in the Three Gorges Reservoir, China. *Lake Reserv. Manag.* **2015**, *31*, 324–338. [CrossRef]
39. Ikeda, S.; Izumi, N. Width and depth of self-formed straight gravel rivers with bank vegetation. *Water Resour. Res.* **1990**, *26*, 2353–2364. [CrossRef]
40. CWRC. *Hydrological data of Changjiang River Basin. Annual Hydrological Report of P. R. China*; Changjiang Water Resources Commission: Beijing, China, 2004.
41. Bridge, J.S. *The Interaction between Channel Geometry, Water Flow, Sediment Transport and Deposition in Braided Rivers*; Geological Society, Special Publications: London, UK, 1993; Volume V75, pp. 13–71.
42. Friend, P.F.; Sinha, R. Braiding and meandering parameters. In *Braided Rivers*; Best, J.L., Bristow, C.S., Eds.; The Geological Society: London, UK, 1993; pp. 105–112.
43. Henderson, F.M. Stability of alluvial channels. *Trans. ASCE* **1963**, *128*, 657–686.
44. Parker, G. Hydraulic geometry of active gravel rivers. *J. Hydraul. Div. ASCE* **1979**, *105*, 1185–1201.
45. Millar, R.G. Theoretical regime equations for mobile gravel-bed rivers with stable banks. *Geomorphology* **2005**, *64*, 207–220. [CrossRef]
46. Eaton, B.C.; Giles, T.R. Assessing the effect of vegetation-related bank strength on channel morphology and stability in gravel bed streams using numerical models. *Earth Surf. Process. Landf.* **2009**, *34*, 712–714. [CrossRef]

Article

Implementation of a Local Time Stepping Algorithm and Its Acceleration Effect on Two-Dimensional Hydrodynamic Models

Xiyan Yang, Wenjie An, Wenda Li and Shanghong Zhang *

Renewable Energy School, North China Electric Power University, Beijing 102206, China; 120192211870@ncepu.edu.cn (X.Y.); 120192211845@ncepu.edu.cn (W.A.); 120192111184@ncepu.edu.cn (W.L.)

* Correspondence: zhangshanghong@ncepu.edu.cn

Received: 11 February 2020; Accepted: 7 April 2020; Published: 17 April 2020

Abstract: The engineering applications of two-dimensional (2D) hydrodynamic models are restricted by the enormous number of meshes needed and the overheads of simulation time. The aim of this study is to improve computational efficiency and optimize the balance between the quantity of grids used in and the simulation accuracy of 2D hydrodynamic models. Local mesh refinement and a local time stepping (LTS) strategy were used to address this aim. The implementation of the LTS algorithm on a 2D shallow-water dynamic model was investigated using the finite volume method on unstructured meshes. The model performance was evaluated using three canonical test cases, which discussed the influential factors and the adaptive conditions of the algorithm. The results of the numerical tests show that the LTS method improved the computational efficiency and fulfilled mass conservation and solution accuracy constraints. Speedup ratios of between 1.3 and 2.1 were obtained. The LTS scheme was used for navigable flow simulation of the river reach between the Three Gorges and Gezhouba Dams. This showed that the LTS scheme is effective for real complex applications and long simulations and can meet the required accuracy. An analysis of the influence of the mesh refinement on the speedup was conducted. Coarse and refined mesh proportions and mesh scales observably affected the acceleration effect of the LTS algorithm. Smaller proportions of refined mesh resulted in higher speedup ratios. Acceleration was the most obvious when mesh scale differences were large. These results provide technical guidelines for reducing computational time for 2D hydrodynamic models on non-uniform unstructured grids.

Keywords: two-dimensional hydrodynamic model; local time stepping; unstructured grids; numerical simulation; computational efficiency

1. Introduction

The numerical simulation of shallow water flow is frequently used in flood forecasting, river regulation, and flood-control planning [1]. Given the need for better accuracy and breadth of shallow-water simulations in engineering applications, a balance between the number of cells, simulation accuracy, and computational time needs to be found. Addressing this issue using optimization of the solving algorithm [2–4] has received much research attention, as has the use of computer hardware acceleration and parallel computing technology [5–7].

The local mesh refinement technique is especially efficient for balancing the number of grids and the accuracy of the simulation. It is flexible and allows grids to be arranged according to different engineering concerns, while retaining a high level of refinement in regions requiring detail, such as where flows change sharply or where there are key features. A coarse mesh is used in general or slow-flowing areas. Consequently, the simulation accuracy can be ensured with only a small increase in the number of grids. This technique has been broadly used in engineering applications involving shallow-water flow simulations [8–12]. However, the time step must strictly satisfy the model solution, which is obtained using the Courant–Friedrichs–Lewy (CFL) stability condition. When using a locally refined mesh, the global time stepping (GTS) size is limited by the time step of the refined mesh, which markedly increases the overall computation time.

In response to this issue, Osher and Sanders [13] proposed a local time stepping (LTS) algorithm for solving the one-dimensional (1D) scalar conservation equations. In each cell, a locally allowable maximum time step satisfying the CFL stability condition was adopted to minimize the computational time. This effectively deals with the complexity of the lag caused by the heterogeneous time step. Tan and Huang [14] developed an efficient adaptive mesh refinement (AMR) with an LTS algorithm for 1D nonlinear hyperbolic problems. Crossley et al. [15,16] successfully applied the LTS technique to the modeling of open channel flows using 1D Saint-Venant equations; their results for unsteady shallow-water equations with LTS met the required accuracy. Consequently, the use of LTS algorithms has enabled significant progress in 1D hydrodynamic simulations. Compared with 1D models for simulating long and wide rivers, two-dimensional (2D) hydrodynamic models are better suited to complex terrains and provide more precise simulation results. Dazzi et al. [17,18] used AMR to implement an LTS algorithm in the numerical simulation of 2D shallow-water flow for a steady-state test case, a circular dam break case, and a Tacker test case. Using the time step of each cell and the CFL condition, they evaluated whether the cell's current time step met the CFL condition. If the condition was not met, then the rule of degrading the time-step rank was implemented, which involved a dynamic inspection of the time step for each cell over the whole computational domain. Wu et al. [19] adopted the LTS technique to avoid excessive storage when modeling three-dimensional (3D) fluid dynamics using the discontinuous finite element method. The powerful function of 3D models determines their complex and elusive performance. At present, hydrodynamic modeling is dominated by 2D modeling, which can provide sufficiently reliable results, such as for river navigation.

Nevertheless, recent research has paid increasing attention to this aspect with a suitable treatment of unstructured grids and has yielded promising results. Although the numerical procedure has unique advantages in terms of mesh generation, storage, use of elements, capturing of specific physical phenomena, and computational efficiency, the boundaries of such regular grids are too generalized and are not geometrically adaptable. Furthermore, the computational accuracy of the boundaries is extremely rough and neglects terrain changes, and is thus limited in its scope. The use of unstructured grids provides highly adaptable solutions for complex terrains. Unstructured grids can incorporate a region's boundaries and constraints using mesh density control in continuous regions. Consequently, unstructured grids play a dominant role in grid generation technology. Using the Runge–Kutta discontinuous Galerkin finite element method, Trahan and Dawson [20] applied LTS to the shallow-water equations and incorporated rainfall, wetting, and drying into the model. Using LTS, Hu et al. [21] established a new coupled model, which had high computational efficiency, for sediment-laden flows. The computational domain was discretized using unstructured triangular meshes. The inter-cell numerical fluxes were estimated using the Harten–Lax–van Leer–Contact (HLLC) approximate Riemann solver. The model used by Hu et al. was applied to the Taipingkou waterway of the Yangtze River and achieved a 92% reduction in computational cost without loss of quantitative accuracy. To solve the problem of mass conservation linked to time splitting in LTS algorithms, Michael [22] improved the finite volume method (FVM) using a deformation correction to achieve local mass conservation, while a row-oriented sub-stepping was proposed to deal with Courant numbers larger than one.

Despite several successful studies based on the LTS algorithm, the computation of flux at the cells with different time-step levels has not been dealt with clearly in recent research. The calculation of the unit variables at the interface is difficult because the use of the local time-step invariably leads to time faults. The calculation of element variables at the interface is crucial but is difficult because of time splitting errors in the LTS algorithm. Implementation of the LTS algorithm on unstructured grids is also valuable for advancing the knowledge base. No specific comparison of the acceleration efficiency of the LTS algorithm has been demonstrated in the existing research. To address these issues, the LTS technique was applied in the establishment of a high-efficiency, 2D, shallow-water, dynamic model using the FVM and unstructured grids. Experiments were performed to verify the flux approximation at the interface, to assess the accuracy of the results, and to analyze the feasibility analysis of the LTS algorithm for complex applications. Model evaluation was performed using 2D simulations of two instantaneous dam break test cases and engineering applications that simulate the navigable flow of the river reach between the Three Gorges and Gezhouba Dams. These tests show the acceleration effect, influencing factors, and the adaptive conditions of the LTS algorithm.

2. Methods

2.1. Global Time Stepping Scheme

2.1.1. Governing Equation

The governing equations used herein are the 2D shallow-water equations (SWEs), written in conservation form [23]:

$$\frac{\partial U}{\partial t}+\frac{\partial F(U)}{\partial x}+\frac{\partial G(U)}{\partial y}=S_0(U,x,y)+S_f(U,x,y)\ (\Omega\times[0,T_s]). \tag{1}$$

In the present study, the modified form of the SWEs was selected to guarantee that the scheme was well-balanced. The variables follow from

$$U=\begin{bmatrix} h \\ hu \\ hv \end{bmatrix}, F(U)=\begin{bmatrix} hu \\ hu^2+\frac{1}{2}gh^2 \\ huv \end{bmatrix}, G(U)=\begin{bmatrix} hv \\ huv \\ hv^2+\frac{1}{2}gh^2 \end{bmatrix},$$
$$S_0=\begin{bmatrix} 0 \\ -gh\frac{\partial z_b}{\partial x} \\ -gh\frac{\partial z_b}{\partial y} \end{bmatrix}, S_f=\begin{bmatrix} 0 \\ -gh\frac{n^2u\sqrt{u^2+v^2}}{h^{4/3}} \\ -gh\frac{n^2v\sqrt{u^2+v^2}}{h^{4/3}} \end{bmatrix},$$

where U is the vector of conserved variables; F and G are the tensors of fluxes in the x and y directions, respectively; S_0 and S_f are the bed and friction slope source terms, respectively; x and y represent the spatial horizontal coordinates; u and v represent the average velocity components from integration of water depth in the x and y directions, respectively; h is the flow depth; z_b is the bed elevation above the datum; g represents the acceleration related to gravity; n is Manning's roughness coefficient; T_s is the duration.

2.1.2. Numerical Technique

Because of the nonlinear characteristics of the SWEs, the analytical solution cannot be obtained generally but can be approximated by numerical discretization. The FVM was used to discretize the governing equation. The computational region was discretized into several points. With these points as the center, the whole computational domain was divided into several interconnected but non-overlapping control volumes. By integrating the basic equation with each control volume, a set of algebraic equations with the unknown quantity on the calculation node was obtained. A triangular mesh was used to discretize the computational domain to adapt the boundary conditions for a complex terrain. The cell-centered mode was used in the calculations, which means that the physical variables are defined in the center of the control volume [24], as shown in Figure 1.

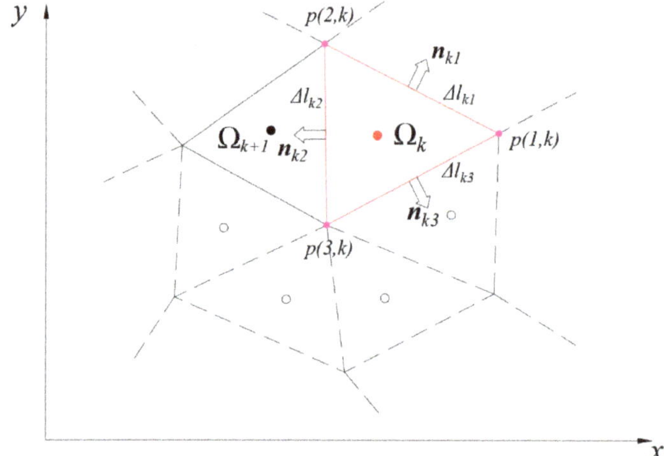

Figure 1. Schematic diagram of the two-dimensional model with finite volume discretization: physical variables are defined at the grid center, where Ω_k is the kth control volume; $\boldsymbol{n}$ is the unit normal vector; $p(i,j)$ is a grid point; and Δl_{kj} is the length of a side.

Integration of Equation (1) over the area of the governing volume Ω, gives

$$\int_{\Omega_i}\left(\frac{\partial U}{\partial t} + \nabla \cdot \mathbf{E}\right) d\Omega = \int_{\Omega_i}\left(S_0 + S_f\right) d\Omega. \tag{2}$$

In Equation (2), $\mathbf{E} = [\mathbf{F}, \mathbf{G}]$ represents a 2D matrix.

The volume integral is transformed to a line integral along the edge of the control volume using Gauss' formula

$$\frac{\Delta u_k}{\Delta t}\Delta s_k = -\int_{l_k} \mathbf{E} \cdot \mathbf{n} dl + \int_{\Omega_k}\left(S_0 + S_f\right) d\Omega, \tag{3}$$

where U_k is the average value of the governing cell; l_k is the edge of the kth control volume Ω_k; Δs_k is the area of each calculation cell; $\mathbf{n}$ is the unit vector of the normal direction outside the edge.

After discretization of Equation (3), the following formula is obtained:

$$\Delta U = -\frac{\Delta t}{\Delta s_k}\sum_{j=1}^{3}\left(\mathbf{E}_{kj}, \mathbf{n}_{kj}\right)\Delta l_{kj} + \frac{\Delta t}{\Delta s_k}\overline{S}, \tag{4}$$

where Δl_{kj} is the length of each side (using triangular grids, $j = 1, 2, 3$); and $\mathbf{E}_{kj}$ and $\mathbf{n}_{kj}$ represent the numerical fluxes and the external normal unit vector of edge j. $\overline{S_0} = \int_{\Omega_i} S_0 d\Omega$ and $\overline{S_f} = \int_{\Omega_i} S_f d\Omega = \Delta s_k S_f$ represent the integral values of the bed and friction slope source terms in the mesh cell, respectively. The notation $\left(\mathbf{E}_{kj}, \mathbf{n}_{kj}\right)$ denotes the inner product.

Commonly used approximate Riemann solvers for the discontinuous problem include the Osher, Roe, Harten–Lax–van Leer (HLL), and HLLC schemes [25]. In the present study, the fluxes at the interface of the governing cell were estimated using the Roe scheme. The physical variables U_L and U_R on the interface of the left and right grid were determined to solve the Riemann discontinuity problem. The flux expression is defined as follows [26–29]:

$$\mathbf{E} \cdot \mathbf{n} = \frac{1}{2}\left[(\mathbf{F}, \mathbf{G})_R \cdot \mathbf{n} + (\mathbf{F}, \mathbf{G})_L \cdot \mathbf{n} - |\widetilde{J}|(U_R - U_L)\right], \tag{5}$$

where $\widetilde{J} = \frac{\partial(\mathbf{E}\cdot\mathbf{n})}{\partial U}$ is the Roe average of the Jacobian matrix. The three eigenvalues of the Jacobian matrix are $\widetilde{\lambda}_k(k = 1, 2, 3)$, while the related eigenvectors are $\widetilde{e}_k$. The left and right interface conservative form variables are U_L and U_R, respectively.

To improve the adaptability of the model to complicated terrain conditions, the source term reflecting the bottom topography was managed using an eigen-decomposition method, while upwind treatment was applied to balance the interface flux [30,31]:

$$\overline{S_0} = \sum_{j=1}^{3}\sum_{k=1}^{3}[\frac{1}{2}(1 - sign(\widetilde{\lambda}_k))\beta_k\widetilde{e}_k l_j]^j, \tag{6}$$

where $sign()$ is the sign function. In this case, $\beta_1 = -\frac{1}{2}\widetilde{c}\widetilde{Z}_b$, $\beta_2 = 0$, and $\beta_3 = \frac{1}{2}\widetilde{c}\widetilde{Z}_b$. Furthermore, $\widetilde{c}$ is the average velocity given by the Roe scheme; and $\widetilde{Z}_b = \frac{1}{2}[(Z_b)_L + (Z_b)_R]$; while $(Z_b)_L$ and $(Z_b)_R$ are the bottom elevations of the left and right side, respectively.

An explicit scheme that discretizes the frictional slope source term was carried out to increase the stability of the computation. Eventually, the variation of the conservation after undergoing Δt was obtained.

$$\Delta U = \mathbf{I}\frac{\Delta t}{A}[(-\sum_{j=1}^{3}\mathbf{E}_j\mathbf{n}_j l_j) + \sum_{j=1}^{3}\sum_{k=1}^{3}(\frac{1}{2}\big(1 - sign(\widetilde{\lambda}_k)\big)\beta_k\widetilde{e}_k l_j)^j + \Delta t S_f^n], \tag{7}$$

where $\mathbf{I}$ is the unit matrix. Superscripts n and $n+1$ refer to the current and updated time levels, respectively. Δt is the globally permissible time step, which must be computed according to the CFL stability condition [32]:

$$\Delta t_i^g = Cr \min_{j=1,2,3}(\frac{d_{i,j}}{\sqrt{u_i^2 + v_i^2} + \sqrt{gh_i}}), \tag{8a}$$

$$\Delta t = \min\left(\Delta t_i^g\right), \tag{8b}$$

where i is the computational cell index, with values of 1, 2, ... , N_e. The term N_e refer to the total number of computational grid elements; $d_{i,j}$ is the distance from the center of the cell (triangle centroid) to the three sides; u_i and v_i are the velocities in the x and y directions, respectively, of cell i; h_i is the flow depth of cell i. The Courant number Cr was set to 0.8.

In the traditional GTS algorithm, the time step Δt is equal to the minimum value for the whole computational domain and is proportional to the size of the grid elements. When the scale of each mesh element varies markedly, the uniform time-step restriction will severely reduce the efficiency of the model operation. Therefore, it is essential to explore an efficient method in which the time-step changes with the mesh size and satisfies the CFL conditions.

2.2. Local Time Stepping Scheme

The key approach in the LTS algorithm is to advance each cell with a time step as close as possible to its maximum allowable value while satisfying the stability of the CFL condition. The algorithm reconstructs the time-step level of each cell with the aim of reducing the computational load of coarse grids, consequently saving computational time. Figure 2 compares the implementation of GTS versus LTS algorithms. The LTS algorithm differs in two aspects: reconstructing the local time step of each cell (Section 2.2.1) and updating values of the conserved variables ΔU (Section 2.2.2).

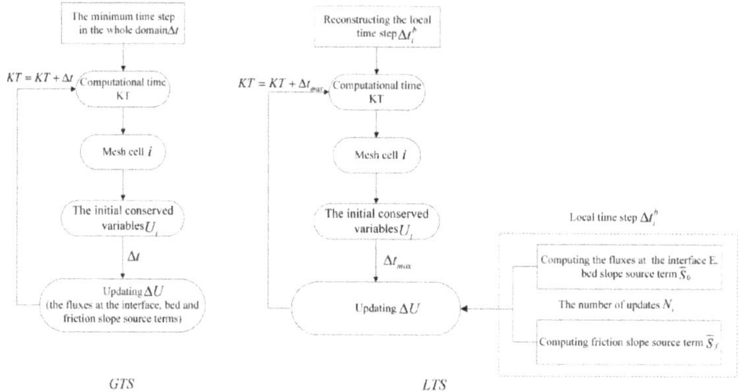

Figure 2. Flow diagram showing global time stepping (GTS) and local time stepping (LTS) algorithms. In the GTS scheme, each cell is updated with the same time step Δt, while in the LTS scheme, a locally allowable maximum time step (Δt_i^b) is used for updating variables in each cell.

2.2.1. Reconstructing the Local Time Step of Each Cell

This procedure is defined as follows:

(1) Compute the initial time step Δt_i^*

The Δt_i^* of each cell satisfying the CFL condition is computed via Equation (9a):

$$\Delta t_i^* = Cr \min_{j=1,2,3} \left(\frac{d_{i,j}}{\sqrt{u^2 + v^2} + \sqrt{gh}} \right), \tag{9a}$$

where u and v are the maximum velocities in the x and y directions, respectively; and h is the maximum water depth.

Next, calculate the minimum time step Δt_r in the whole computational domain:

$$\Delta t_r = \min\left(\Delta t_i^*\right). \tag{9b}$$

(2) Set an initial temporal resolution level for each cell

In terms of the CFL condition, the initial time step Δt_i^* in cell i is proportional to the mesh scale $d_{i,j}$. Taking the minimum time step Δt_r over the whole computational domain as the reference standard, a binary logarithmic function is selected as the conversion tool. Each cell is classified based on its initial time step Δt_i^*, with the allocation of an initial time-step level m_i^*, reflecting the mesh scale and having integer values of 1, 2, 4, 8, 16, 32, 64, ... The initial m-level distribution map is shown in Figure 3a.

$$m_i^* = \text{int}(\log_2\left(\frac{\Delta t_i^*}{\Delta t_r}\right)), \tag{10}$$

where the function of $int()$ is to obtain the maximum integer that does not exceed the given number.

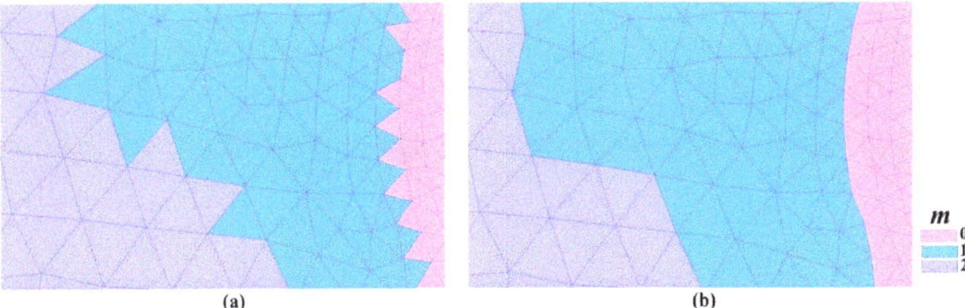

Figure 3. Example of the m-level assignment to each cell in the unstructured grid. (**a**) Distribution showing initial values assigned by Equation (10); (**b**) distribution after being modified by Equation (11a), with the reassignment of buffer zone values based on neighboring cells values.

(3) Modify the time-step level

The initial m-level distribution calculated by Equation (10) is usually scattered and broken (see Figure 3a), especially within the transition zone between coarse and refined grids. This distribution varies dramatically because of the large difference in mesh size. Appropriate measures need to be taken to smooth the transition and to reduce mass and momentum errors in these zones related to the calculation of conserved variables at interfaces between cells with different m-levels. The initial m-levels are modified via Equation (11a), which produces a buffer zone.

$$m_i = \min\left(m_i^*,\ m_{i,1},\ m_{i,2},\ m_{i,3}\right), \tag{11a}$$

where $m_{i,j}(\mathrm{j} = 1,2,3)$ is the time-step level of the three neighboring cells of cell i. The evaluation of the time-step level m_i in cell i is based on values of the neighboring cells: $m_{i,1},\ m_{i,2},\ m_{i,3}$. If the time-step level of the current cell is larger than that of the neighboring cells, then the m_i value will be reduced by one; otherwise, it will be retained. Thus, the difference between the m-levels of the current and neighboring cells cannot exceed one. In this way, the time-step level m_i (see Figure 3a) is modified via Equation (11a), and the new m-level distribution is as shown in Figure 3b.

The maximum of time-step level m_{max} is given by

$$m_{max} = \max(m_i), \tag{11b}$$

where $L = m_{max} + 1$ refers to the local time-step rank of the LTS algorithm in the computation (L = 1 represents the GTS solution). Building on the requirement of simulation accuracy, the maximum value of the different time-step level m_u is selected according to the simulation accuracy requirements, and the restriction condition $m_i = \min\left(m_u,\ m_i^*,\ m_{i,1},\ m_{i,2},\ m_{i,3}\right)$ is added to Equation (11a).

In this case, the time step adopted by each calculation cell Δt_i^b is

$$\Delta t_i^b = 2^{m_i} \Delta t_r. \tag{11c}$$

The maximum time step Δt_{max} is

$$\Delta t_{max} = 2^{m_{max}} \Delta t_r \tag{11d}$$

2.2.2. Calculation of Element Variables at the Interface

Once the local time-step level of each computation cell is reconstructed, no further cells are updated with the same time step but are instead updated with the locally allowable time step at each cell. The algorithm focuses on the convergence of the element variables at different time levels along the interface. In a maximum time step Δt_{max}, the number of updates at each cell N_i is calculated using

$$N_i = 2^{m_{max}-m_i}. \tag{12}$$

The expression for ΔU in Equation (7) is then replaced by Equation (13), in which Δt_i^b from Equation (11c) takes the place of Δt_r in each computing cell. The improved expression is as follows:

$$\Delta U = \mathbf{I}\frac{\Delta t_i^b}{A}[(-\sum_{j=1}^{3}\mathbf{E}_j\mathbf{n}_j l_j) + \sum_{j=1}^{3}\sum_{k=1}^{3}(\frac{1}{2}\big(1-sign(\widetilde{\lambda}_k)\big)\beta_k\widetilde{e}_k l_j)^j + \Delta t_i^b S_f^n]. \tag{13}$$

Then, Equation (13) is iterated N_i times, and the conserved variables of each cell are updated to the same time level n, from which the variables at the next time level $n+1$ are calculated until the calculation time terminates.

The fluxes on the interface of the governing cell are estimated from the approximate Riemann solution of the Roe scheme. However, the variable values of all cells at the current time level need to be given to confirm the U_R and U_L values of the adjacent states on the left and right sides of the interface. In the initial GTS model, the unified global time step Δt can meet these requirements. In the LTS algorithm, however, when the m-level of the current and neighboring cells differ within the maximum time step Δt_{max}, the update times are different, and time splitting occurs at the interface. In the present study, the value of the missing element variable at the interface was obtained by linear interpolation of the value of the element variable in two adjacent time layers. This is shown in Figure 4, where the maximum time-step level m_{max} = 2. At the starting time of 0, we assumed that Δt_r = 0.01 s. First, the element variables of the coarse grid (m = 2) were computed only once, which consumes 0.04 s. Next, the values for the intermediate meshes (m = 1) were calculated twice, costing 0.02 s for each calculation. The local time step of the coarse mesh (m = 2) at the interface is, however, 0.04 s, which suggests that the variable values at 0.02 s are missing. In the present study, the variable values $U(h, u, v)$ were estimated approximately using the average value of the former and the latter time step (1/2 of the variable value at 0 and 0.04 s). Finally, the refined grids (m = 0) were calculated at four intervals, with time steps of 0.01 s. At 0.01 and 0.03 s, the values of the variables are absent at the interface of the cells having m = 1. The above approximation method was again used. At 0.04 s, all cell variables were updated to the same time level in preparation for the next iteration.

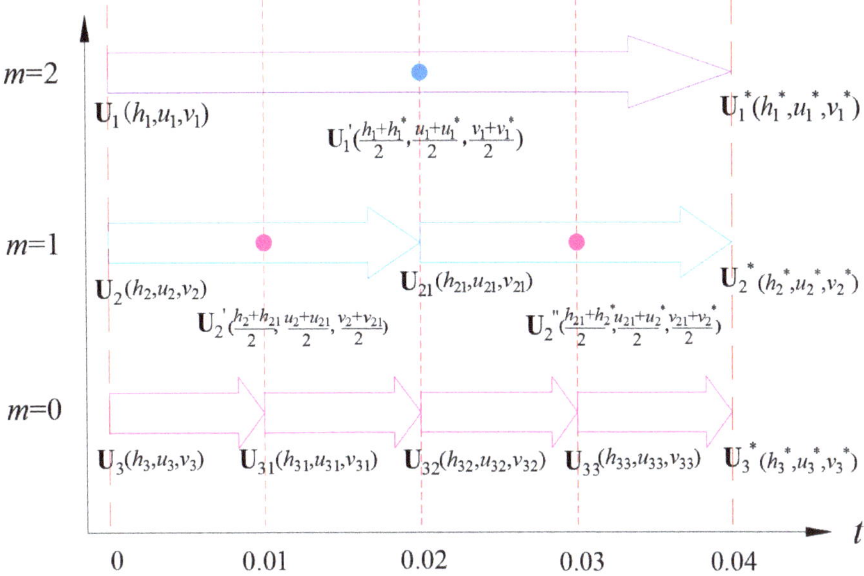

Figure 4. Sketch showing the calculation of element variables at an interface: the variable $U(h,u,v)$ refers to the water depth h, and the velocity components u and v in the x and y directions, respectively. Within a maximum time step of $\Delta t_{max} = 0.04$ s, the coarse mesh areas (time-step level, $m = 2$) were updated only once. The value of $U_1(h_1,u_1,v_1)$ at 0 s was updated to $U_1^*(h_1^*,u_1^*,v_1^*)$ after 0.04 s. The blank value at 0.02 s was estimated using the average of the values at the former and the latter time-step level, giving $U_1'\left(\frac{h_1+h_1^*}{2},\frac{u_1+u_1^*}{2},\frac{v_1+v_1^*}{2}\right)$. Immediately following this, the intermediate grids ($m = 1$) were updated twice over this maximum time step. The values of variables at 0.01 and 0.03 s were similarly approximated. Finally, the refined mesh grids ($m = 0$) were updated four times over this interval. All variables were ultimately updated to 0.04 s, which allowed the next time step to be performed.

3. Numerical Tests

Numerical tests were carried out to verify the reliability and adaptability of the LTS strategy within a 2D hydrodynamic model, and to make a comparison between the effects of the novel LTS and the original GTS strategies on the simulation results. To quantify these differences, we used the mean-square error (MSE)

$$L_s(f) = \sqrt{\frac{1}{N_e}\sum_{i=1}^{N_e}(f_{i,GTS} - f_{i,L})^2}, \tag{14}$$

where $L_s(f)$ refers to the MSE of variables f (such as water depth h, and velocity components u and v in the x and y directions, respectively), and the subscript L refers to the local time-step rank of the solutions.

The principle purpose of using the LTS algorithm instead of the GTS algorithm is to reduce the computing time. The relative time saving ratio T_r and speedup ratio S_n were used as evaluation indices of the computing efficiency for different L values.

The time saving ratio T_r is given by

$$T_r = \frac{T_{GTS} - T_L}{T_{GTS}} \times 100. \tag{15a}$$

The speedup ratio S_n is given by

$$S_n = \frac{T_{GTS}}{T_L}, \tag{15b}$$

where T_{GTS} is the computation time of the GTS strategy, and T_L is the computation time of the LTS strategy using different local time-step rank values (L).

3.1. Anti-Symmetric Dam Break Case

A classic verification experiment that considers the anti-symmetric square dam break case has been described by Fennema and Chaudhry [33]. The computational domain consists of a 200 × 200 m square region with a flat bottom (bed elevation of 0 m). In this problem, the dam is assumed to fail instantaneously at a certain moment, producing an anti-symmetric breach with a width of 75 m at the center of the calculation domain, which is 95 m from the x-axis boundary. The detailed geometry of this problem is shown in Figure 5.

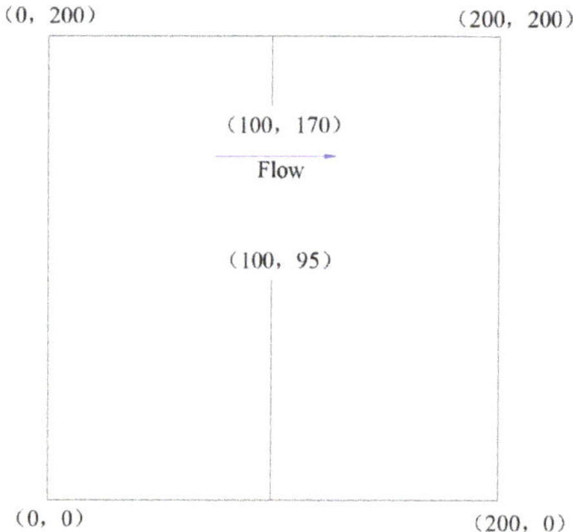

Figure 5. Sketch showing the plane geometry of an anti-symmetric dam break, with the breach occurring in the center.

The computational domain was discretized on triangular meshes. Local mesh refinement technology was adopted near the break to accurately reflect the important flow regions while controlling the number of grids. The spatial resolution level varied from 6 to 1 m, with a refined area of 5%. A total of 13,338 computing cells were generated. The LTS algorithm was applied to shorten the computing time and improve the calculation efficiency. The m-level distribution of the LTS strategy is presented in Figure 6. The simulation duration was T_s = 160 s. At the initial moment, upstream water depth was set to 10.0 m, while downstream water depth was assumed to be 5.0 m. Manning's roughness coefficient was set to 0.03 [34].

Before making a quantitative analysis of the MSE generated by the LTS strategy, it is useful to verify the validity of the GTS scheme applied to the original model. The numerical solution results at t = 7.2 s were obtained using the GTS strategy; the resulting water surface profiles are shown in Figure 7a, contours of water depth in Figure 7b, and velocity vector distribution in Figure 7c. These results effectively simulated the dam break waves and are consistent with previous research results [34–36].

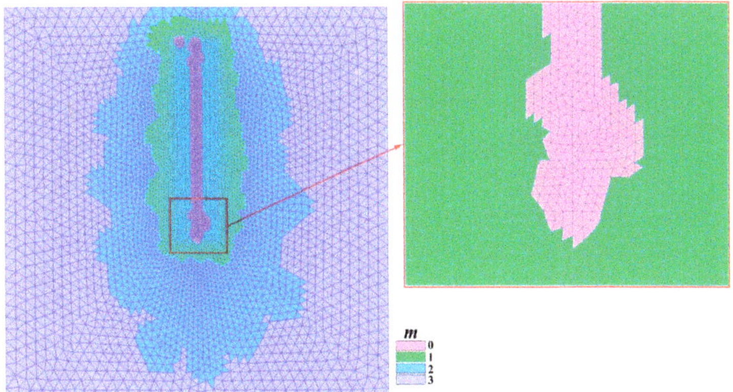

Figure 6. Details of the non-uniform unstructured meshes and time-step level (m-level) distribution for the anti-symmetric dam break case.

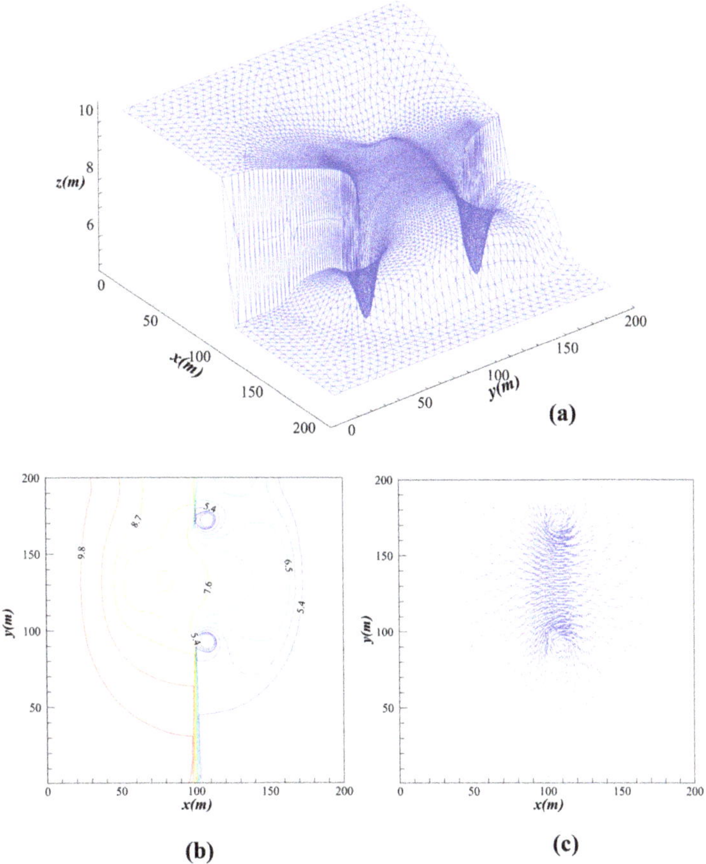

Figure 7. Results of the simulation using the global time stepping strategy at dam break (t = 7.2 s). (**a**) Three-dimensional visualization of the water surface; (**b**) contour plot of water depth; (**c**) velocity vector distribution.

Table 1 shows the MSE of water depth h, as well as velocity components u and v along the x and y directions, respectively, of the simulation results. As the value of the local time-step rank L increased, the MSE increased accordingly. Both precision and efficiency are important when performing hydrodynamic numerical simulations. By reducing the value of L, the calculation efficiency can be improved when levels of high accuracy are reached. This observation agrees well with the research results of Hu et al. [37]. In the present study, the error of the solution using the LTS algorithm was within the 10^{-3} to 10^{-2} range. At the beginning of the calculation, the total water volume was 3×10^5 m^3. When the value of L was 2, 3, and 4, the relative error of the water volume was 0.0017%, 0.0019%, and 0.0045%, respectively, which indicates that the simulation meets the accuracy and conservation of water volume requirements.

Table 2 shows the computing time, time-saving ratio, and speedup ratio for different values of L. The simulation time of the dam break was 160 s. A time saving of 40% to 50% was achieved using the LTS algorithm in the numerical experiment of the anti-symmetric square dam break. For L = 4, a speedup ratio of 2.1 was achieved.

This clearly shows that the calculation efficiency of the numerical simulation of the anti-symmetric square dam break is improved using the LTS technique (Figure 8). As the value of L increased, the speedup ratio also increased, thus shortening the model operation.

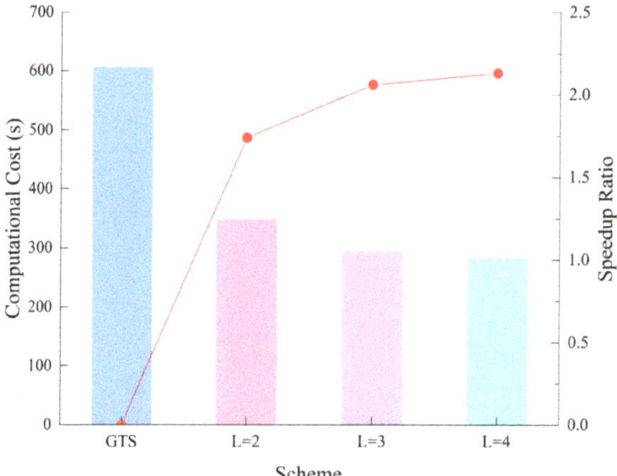

Figure 8. Computational cost (s) and speedup ratios for different local time-step rank values (L) for the anti-symmetric dam break case. As the value of L increases, the computational cost decreases and the speedup ratio increases.

Table 1. Mean square errors of water depth (h) and velocity components (u, v) for different local time-step rank values (L) in the anti-symmetric dam break case.

t(s)	L = 2			L = 3			L = 4		
	$L_s(u) \times 10^{-2}$	$L_s(v) \times 10^{-2}$	$L_s(h) \times 10^{-2}$	$L_s(u) \times 10^{-2}$	$L_s(v) \times 10^{-2}$	$L_s(h) \times 10^{-2}$	$L_s(u) \times 10^{-2}$	$L_s(v) \times 10^{-2}$	$L_s(h) \times 10^{-2}$
7.2	0.41	0.38	0.22	0.65	0.56	0.50	1.07	0.71	0.79
15.2	0.34	0.31	0.17	0.65	0.64	0.45	1.02	0.67	0.87
23.2	0.50	0.43	0.39	1.02	0.81	1.07	1.84	1.20	1.99
31.2	0.66	0.72	0.57	1.53	1.45	1.39	2.81	2.63	2.62
39.2	0.74	0.75	0.29	1.34	1.37	0.62	2.43	2.40	1.04
47.2	0.55	0.70	0.30	0.89	1.24	0.72	2.03	3.31	1.63
55.2	0.55	0.71	0.36	0.84	1.06	0.96	1.72	2.16	1.76
63.2	0.51	0.60	0.31	0.95	1.18	0.86	1.63	3.12	2.14
71.2	0.71	0.70	0.30	1.17	1.25	0.68	2.82	2.81	1.46
79.2	0.64	0.63	0.27	1.10	1.05	0.74	2.54	2.32	1.50
120	0.67	0.63	0.18	1.33	1.12	0.47	2.69	2.26	1.10
160	0.61	0.56	0.24	1.03	0.97	0.70	2.35	2.26	1.85
average	0.57	0.59	0.30	1.04	1.06	0.76	2.08	2.15	1.56

Table 2. Computing time T (s), time-saving ratio T_r (%), speedup ratio (S_n) of different local time-step rank values (L) for the anti-symmetric dam break case (GTS: global time stepping strategy).

Test	T	T_r	S_n
GTS	608	-	-
L = 2	349	43	1.74
L = 3	295	51	2.06
L = 4	285	53	2.13

3.2. Non-Flat Bottom Dam Break Case

In this test case, the length of the rectangular channel was 75 m, and its width was 30 m. Three obstacles were placed on the bottom of the rectangular channel. Two obstacles had a radius of 6.5 m and a height of 1 m, and one had a radius of 8.6 m and a height of 2 m. The elevation expression is defined as follows:

$$Z_b(x,y) = \begin{cases} 1 - [(x-30)^2 + (y-8)^2]/42.25, & (x-30)^2 + (y-8)^2 \leq 42.25 \\ 1 - [(x-30)^2 + (y-22)^2]/42.25, & (x-30)^2 + (y-22)^2 \leq 42.25 \\ 2 - [(x-52)^2 + (y-15)^2]/36.89, & (x-52)^2 + (y-15)^2 \leq 73.98 \\ 0, \text{ else} \end{cases}.$$

Local mesh refinement was applied with a spatial resolution level varying from 0.5 to 3 m. The grid generation results are shown in Figure 9. The total number of grid cells for the region was 6848. In the calculation, a solid wall boundary condition was adopted. Manning's roughness coefficient was set to 0.018 [30]. The water depth for $x \leq 16$ m in the computational domain was set to 2 m and was assumed to be 0 m elsewhere. The variation of the ground level is shown in Figure 10a. In the GTS scheme, the time step was 0.005 s, the initial total water volume was 959.841 m^3, and the final water volume was 959.978 m^3. The relative error was 0.01%, which is within the discrete error range. The model dealt well with wetting and drying fronts over the complex terrain, satisfied computational stability, and conservation of water volume requirements, and provided results that were consistent with previous experimental results [38,39].

Figure 9. Details of the non-uniform unstructured meshes, time-step level (m-level) distribution, and bathymetric point arrangement for the non-flat bottom dam break case.

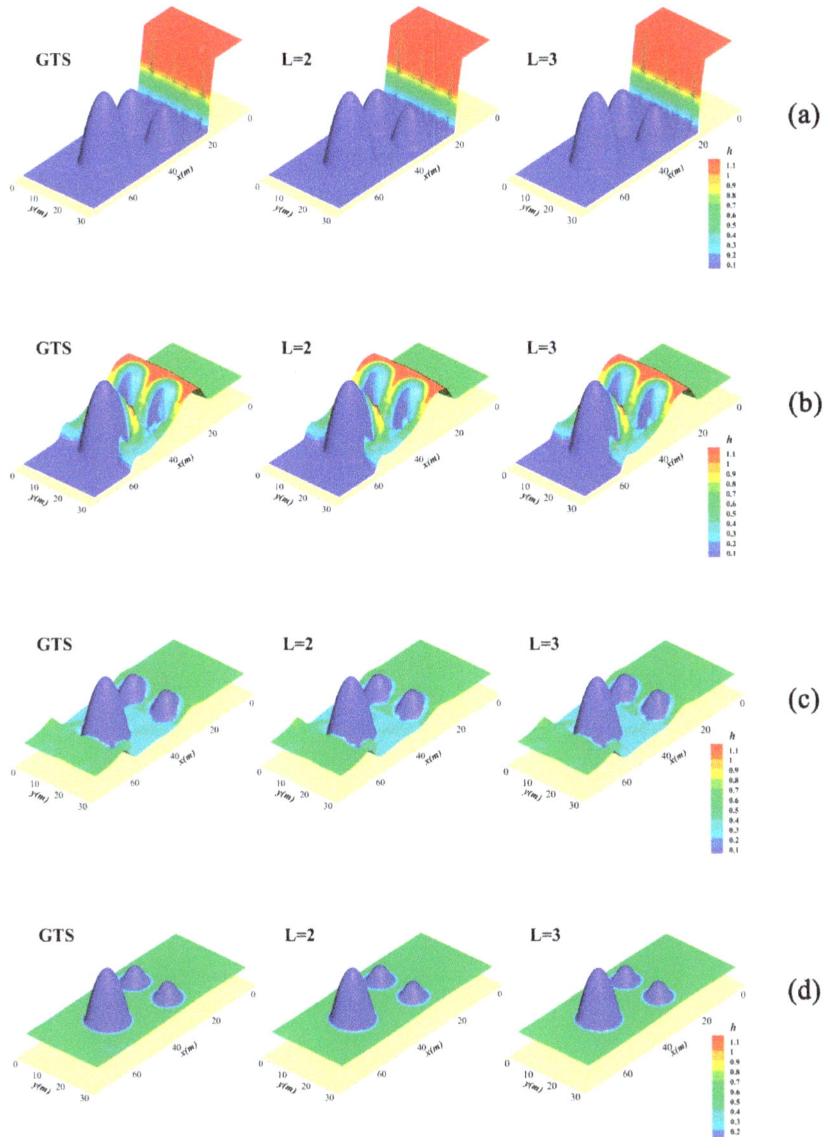

Figure 10. Three-dimensional visualization of the inundation process at various time points following the dam break. (**a**) t = 0 s; (**b**) t = 10 s; (**c**) t = 30 s; (**d**) t = 100 s. (GTS: global time stepping strategy, while L is the time-step rank value in the local time stepping strategy).

The distribution of m-values obtained by applying the LTS strategy is shown in Figure 9. The maximum time-step level m_{max} is 2. The simulation of the dam break water flow was modeled for T_s = 100 s. The inundation related to the dam break is shown in Figure 10. To observe the influence of the LTS algorithm on the simulation results, four bathymetric measuring points were arranged (see Figure 9 for the location of the measuring points). Figure 11 provides a comparison of water depths generated by the LTS algorithm and the classical algorithm during the process of the dam break. A visual inspection indicates that the simulated water depths from the two algorithms are essentially the same. Thus, the LTS technique improves the efficiency of calculation without any reduction in accuracy.

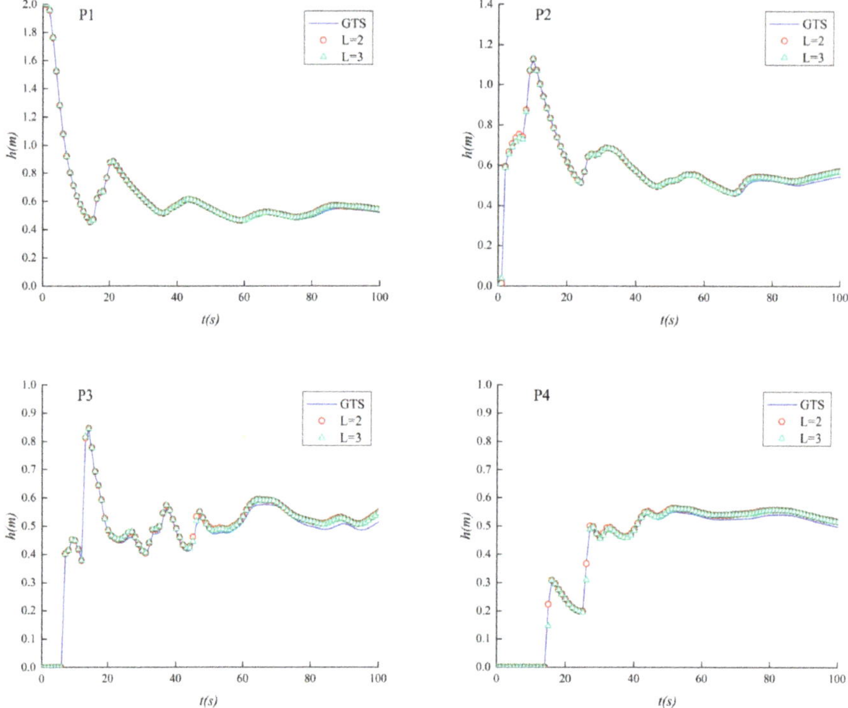

Figure 11. Comparison of water levels at different local time-step rank values (L) for the non-flat bottom dam break case (GTS: global time stepping strategy).

According to the statistical results for this simulation (shown in Table 3), adopting the LTS technology in the numerical experiment of the non-flat bottom dam break created a saving of up to 26% in computation time, and a speedup ratio of 1.3, without compromising the solution accuracy. The best result was obtained for L = 3, which reduced the computational cost and improved the calculation efficiency.

Table 3. Computing time T (s), time saving ratio T_r (%), speedup ratio (S_n) for different local time-step rank values (L) for the non-flat bottom dam break case (GTS: global time stepping strategy).

Test	T	T_r	S_n
GTS	148	-	-
L = 2	115	22	1.28
L = 3	110	26	1.34

3.3. Navigable Flow Simulation Case

There are several noticeable perilous waterways between the Three Gorges Dam and the Gezhouba Dam. The river section is about 38 km long [40,41]. Bed elevations measured in October 2008 provided by the Three Gorges Navigation Authority were used as the initial bed topography. In the present study, the three segments with the highest risk to shipping (the Letianxi, Liantuo, and Shipai riverways), and the outlet of the Three Gorges Reservoir are adopted for local mesh refinement. The spatial resolution level ranged from 10 to 100 m, and 36,297 cells and 19,011 nodes were generated. The results of the grid generation and refined areas are shown in Figure 12.

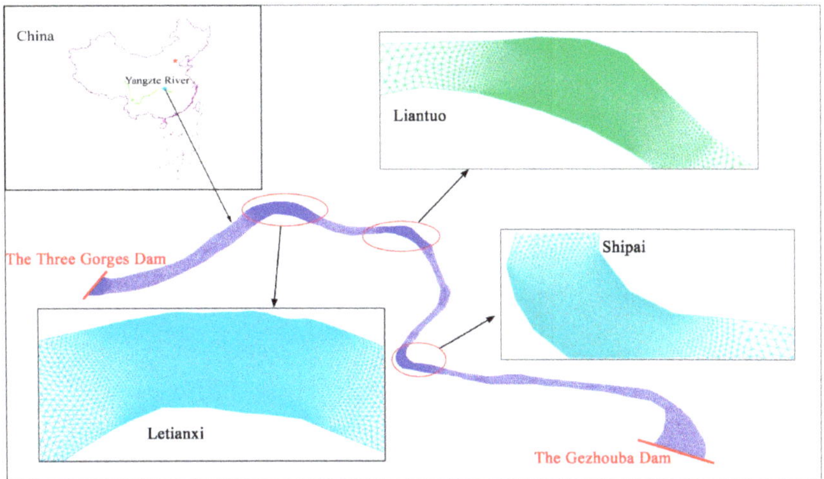

Figure 12. Computational cells and details of refined areas for the river reach between the Three Gorges and Gezhouba Dams.

The LTS algorithm was applied to the numerical simulation of the navigable flow of the waterway. The results are provided in Table 4. The maximum time-step level (m_{max} = 4) and the minimum time step (Δt_r = 0.03 s) were calculated for the whole computational domain. The simulation duration was set to T_s = 4 h. The absolute deviation in the water level between the calculated and measured values was 0.06 to 0.07 m, and the relative deviation in flow velocity was between −3.8% and 4.6%. There is no significant difference between the LTS and GTS algorithm simulation results for water level and flow velocity. The mass errors are non-negligible for a large computational area and a long simulation. This was addressed by adopting a smaller maximum time-step level m_{max}. The obvious acceleration effect that this provides in the real application ensures that accuracy requirements are met. It is clear that the LTS algorithm has provided obvious improvements in the execution time without compromising the solution accuracy. This is of practical value and importance in navigation.

Table 4. Simulation results for the navigable flow of the waterway between the Three Gorges and Gezhouba Dams.

Data		**River Reach**		**Water Level (m)**						**Flow Velocity (m/s)**					
Observation Time	**Discharge (m^3/s)**			**Measured**	**Calculated**					**Measured**	**Calculated**				
					GTS	**L = 2**	**L = 3**	**L = 4**	**L = 5**		**GTS**	**L = 2**	**L = 3**	**L = 4**	**L = 5**
2008.08.20	28,400	Letianxi	Value	68.32	68.39	68.39	68.39	68.39	68.39	1.56	1.63	1.63	1.63	1.63	1.63
			Deviation	-	0.07	0.07	0.07	0.07	0.07	-	0.07	0.07	0.07	0.07	0.07
2008.08.20	29,700	Liantuo	Value	68.35	68.41	68.41	68.41	68.41	68.41	2.13	2.16	2.16	2.16	2.16	2.16
			Deviation	-	0.06	0.06	0.06	0.06	0.06	-	0.03	0.03	0.03	0.03	0.03
2008.09.04	31,700	Shipai	Value	67.83	67.90	67.90	67.90	67.90	67.90	1.61	1.55	1.55	1.55	1.55	1.55
			Deviation	-	0.07	0.07	0.07	0.07	0.07	-	−0.06	−0.06	−0.06	−0.06	−0.06
		T			6.65	3.86	3.56	3.45	3.39						
		T_r			-	41.96	46.47	48.06	48.96						
		S_n			-	1.72	1.87	1.93	1.96						

GTS: global time stepping strategy; L: local time-step rank in the local time stepping strategy; T: execution time (*h*); T_r: time saving ratio (%); and S_n: speedup ratio.

4. Discussion

4.1. The Influence of the Proportion of Refined Mesh on the Acceleration Effect

In the anti-symmetric dam break case, the 75-m breach was used as the core area in which to analyze the acceleration effect of the LTS algorithm. In this area, we set the refined area to different proportions of the total area (4 hm^2). The spatial resolution level varied from 1 to 6 m, with a minimum resolution close to the breach of 1 m, surrounded by a smooth transition up to a maximum resolution of 6 m in the outer domain. The mesh generation results for the refined areas for various proportions in the anti-symmetric dam break case is shown in Figure 13. The numerical simulation was performed for refined areas of 5%, 10%, 25%, 50%, and 75%. The impact of increasing the local refinement area on the acceleration effect was analyzed for a time-step level of $m = 0$ in the refinement area, $m = 2$ in the coarse grid area, and $m = 1$ in the buffer zone.

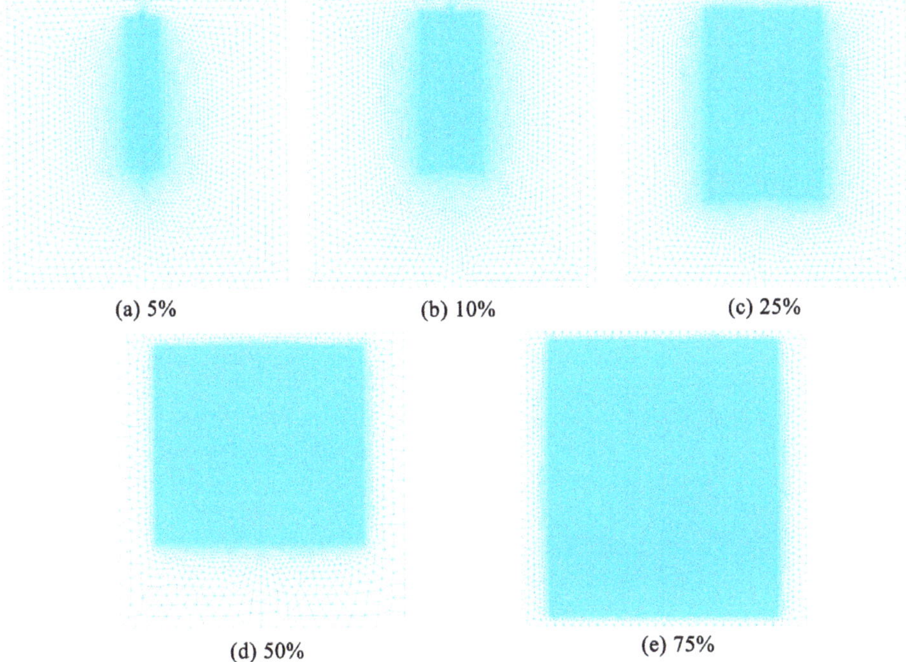

Figure 13. Locally refined areas of various proportions (%) in the anti-symmetric dam break case: the refined area of (**a**) 5%; (**b**) 10%; (**c**) 25%; (**d**) 50%, (**e**) 75%.

The duration of the dam break T_s was kept at 160 s. Table 5 shows the number of generated grids, computational cost, speedup ratio, and time-saving ratio for different refined area proportions. As the refined area was increased, the total number of meshes generated by partitioning increased. The grid number increased from 13,338 at 0.2 hm^2 to 73,306 at 3 hm^2, which represents an increase by a factor of 5.5. The operating time of the model rose from 608 to 3394 s, which represents an increase by a factor of 5.6. The number of grids is a key factor determining the overhead of computing. There is a direct positive correlation between the number of grids and computing overheads.

Table 5. Simulation results for different refined area proportions in the dam break simulation.

Refined Area Proportion	Refined Area (hm^2)	Total Grid Number	Test			
			Index	GTS	L = 2	L = 3
5%	0.2	13,338	T	608	467	436
			T_r	-	23%	28%
			S_n	-	1.30	1.39
10%	0.4	18,986	T	857	727	702
			T_r	-	15%	18%
			S_n	-	1.18	1.22
25%	1	32,230	T	1474	1328	1308
			T_r	-	9.8%	11.3%
			S_n	-	1.11	1.13
50%	2	48,622	T	2234	2197	2183
			T_r	-	1.6%	2.3%
			S_n	-	1.01	1.02
75%	3	73,306	T	3393	3376	3352
			T_r	-	0.5%	0.5%
			S_n	-	1.01	1.01

NB: GTS: global time stepping strategy; L: local time-step rank in the local time stepping strategy; T: execution time (s), T_r: time-saving ratio, S_n: speedup ratio.

The speedup ratio was strongly affected by the ratio of the locally refined area to the total area. As the proportion of the refined area increased, the speedup ratio obtained by the LTS algorithm diminished, reducing the acceleration effect. When L = 3, the speedup ratio decreased from 1.39 times to 1.01 times; when the refined area was increased from 5% to 75%, there was scarcely any improvement in computational efficiency at high proportions of refined area. These patterns occurred because a high proportion of refined area brings about an unreasonable number of refined grids (with a local time-step level, $m = 0$), which reduces the number of coarse grids ($m = 1, 2$) that help optimize the operation. This suggests that little improvement is gained by implementing the LTS algorithm in cases that have a high proportion of refined grids. Conversely, when refined grids account for a small proportion of the total number of grids, the LTS algorithm can markedly shorten the computing time and achieve an excellent acceleration effect. Furthermore, the results in Figure 14 suggest that L has an influence on the calculation efficiency. The acceleration ratio increased as L increased.

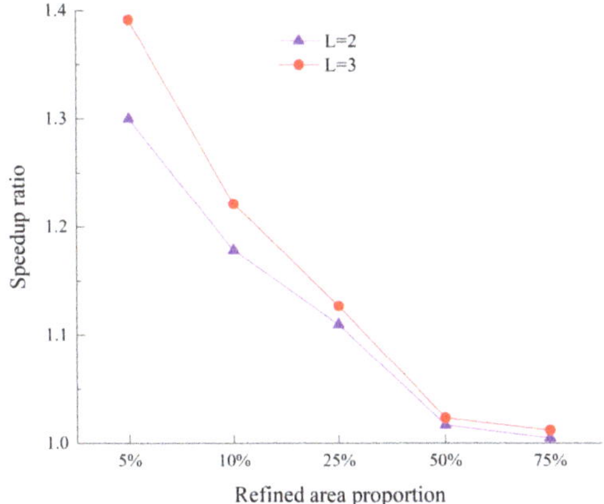

Figure 14. Speedup ratio for implementing the LTS algorithm for various refined area proportions in the dam break model for different local time-step rank values (L). As the refined area increased, the speedup ratio decreased. The acceleration effect increased with increasing L.

4.2. The Impact of the Different Scale of the Mesh on Acceleration Effect

In the case study of the anti-symmetric dam break, the test was carried out using a refinement of 5% of the breach area, as shown in Figure 13a. A grid discretization method was employed to analyze the influence of using different grid scales on the acceleration effect while using the LTS algorithm. The refined grid had a minimum spatial step of 1 m in the burst section, while the coarse grids had spatial steps of 2, 3, 4, 5, and 6 m. The computational results are shown in Table 6 and Figure 15. These results indicate that the speedup ratio improves with the increasing differences in grid resolution. The CFL condition of Equation (9a) shows that the time step of the mesh is positively correlated with the mesh scale difference. As the difference between the maximum and minimum scales of the grids increases, the maximum time-step level m_{max} becomes higher, and a larger L can be adopted. In this test, once the spatial step of the coarse grid was 4, 5, and 6 m, it was possible to set the L value to 3, while the other two conditions were only subdivided into two grades. This explains why greater grid diversity leads to a higher speedup ratio.

Table 6. Simulation results for different spatial resolutions in the dam break simulation.

Spatial Resolution	Grid Number	Test		
		Index	GTS	LTS
1–2 m	35,104	T	1558	1223
		T_r	-	22%
		S_n	-	1.27
1–3 m	22,084	T	973	769
		T_r	-	21%
		S_n	-	1.26
1–4 m	17,108	T	757	589
		T_r	-	22%
		S_n	-	1.28
1–5 m	14,380	T	641	494
		T_r	-	23%
		S_n	-	1.30
1–6 m	13,338	T	608	436
		T_r	-	28%
		S_n	-	1.39

(GTS: global time stepping strategy, LTS: local time stepping strategy, T: execution time (s), T_r: time-saving ratio, and S_n: speedup ratio).

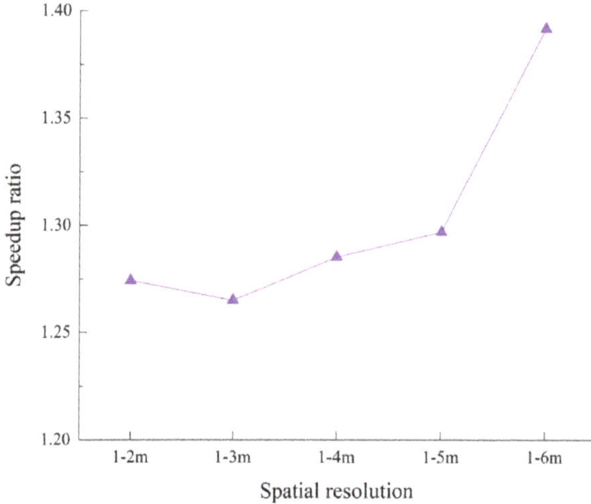

Figure 15. Graph showing the relationship between speedup ratio and grid scale difference.

5. Conclusions

Improvements in the computational efficiency of the 2D hydrodynamic model have a significant bearing on its engineering applications. In the present study, an improved 2D shallow-water dynamic model applying an LTS algorithm was established using an FVM on a triangular mesh. Results from simulation and analysis of three canonical test cases show that the LTS scheme has a satisfactory ability for adapting real complex applications and long simulations while meeting the required accuracy. The following conclusions were drawn:

(1) Based on the FVM for unstructured grids, an LTS algorithm was implemented that improved the computational efficiency of the model, while satisfying water conservation conditions. In the anti-symmetric dam break case, a speedup ratio of 2.1 was achieved, which saved 53% in execution time. The speedup ratio of the non-flat bottom dam break case was 1.3, which represented a shortening of 26% in the calculation time. The numerical simulation of the navigable flow of the river reach between the Three Gorges and Gezhouba Dams achieved a speedup ratio of 1.9, which represented a saving of 49% in modeling time.

(2) The proportions of coarse to refined meshes on the acceleration effect of the LTS algorithm were noticeable. It was evident that a higher speedup ratio was obtained when the proportion of the refined mesh was minimized. When the proportion of the refined mesh was high, the acceleration effect was not significant. It is clear that the LTS algorithm is best suited to situations in which refinement is only required in small regions.

(3) When using the LTS algorithm on non-uniform unstructured grids, the larger the grid scale difference, the more obvious the grid layering became. This led to increased acceleration effects. However, computational accuracy was slightly impaired by excessive differences in grid mesh size.

Author Contributions: Conceptualization, S.Z. and X.Y.; Methodology, S.Z. and X.Y.; Software, X.Y., W.A. and W.L.; Validation, S.Z., X.Y., W.A. and W.L.; Formal Analysis, S.Z.; Investigation, X.Y. and W.A.; Resources, S.Z.; Data Curation, X.Y. and W.A.; Writing—Original Draft Preparation, X.Y.; Writing—Review & Editing, S.Z.; Supervision, S.Z. All authors have read and agreed to the published version of the manuscript.

Funding: This study was supported by the National Natural Science Foundation of China (51979105, 51722901).

Acknowledgments: The authors thank Trudi Semeniuk and Paul Seward from Liwen Bianji, Edanz Editing China (www.liwenbianji.cn/ac), for editing the English text of drafts of this manuscript.

Conflicts of Interest: The authors declare no conflict of interest.

References

1. Wang, Z.; Gang, Y.; Jin, S. Numerical modeling of 2D shallow water flow with complicated geometry and topography. *J. Hydraul. Eng.* **2005**, *36*, 439–444. (In Chinese)
2. Jin, H.; Jespersen, D.; Mehrotra, P.; Biswas, R.; Huang, L.; Chapman, B. High performance computing using MPI and OpenMP on multi-core parallel systems. *Parallel Comput.* **2011**, *37*, 562–575. [CrossRef]
3. Zhang, S.; Xia, Z.; Yuan, R.; Jiang, X. Parallel computation of a dam-break flow model using OpenMP on a multi-core computer. *J. Hydrol.* **2014**, *512*, 126–133. [CrossRef]
4. Vacondio, R.; Dal Palù, A.; Ferrari, A.; Mignosa, P.; Aureli, F.; Dazzi, S. A non-uniform efficient grid type for GPU-parallel Shallow Water Equations models. *Environ. Model. Softw.* **2017**, *88*, 119–137. [CrossRef]
5. Dawson, C. High Resolution Schemes For Conservation Laws With Locally Varying Time Steps. *SIAM J. Sci. Comput.* **2000**, *22*, 2256–2281. [CrossRef]
6. Sanders, B.F. Integration of a shallow water model with a local time step. *J. Hydraul. Res.* **2008**, *46*, 466–475. [CrossRef]
7. Kesserwani, G.; Liang, Q. RKDG2 shallow-water solver on non-uniform grids with local time steps: Application to 1D and 2D hydrodynamics. *Appl. Math. Model.* **2015**, *39*, 1317–1340. [CrossRef]
8. Caboussat, A.; Boyaval, S.; Masserey, A. On the modeling and simulation of non-hydrostatic dam break flows. *Comput. Vis. Sci.* **2011**, *14*, 401–417. [CrossRef]
9. Li, D.; Zeng, Q.; Feng, H. A P-adaptive Discontinuous Galerkin Method Using Local Time-stepping Strategy Applied to the Shallow Water Equations. *J. Inf. Comput. Sci.* **2013**, *10*, 2199–2210. [CrossRef]
10. Tirupathi, S.; Hesthaven, J.S.; Liang, Y.; Parmentier, M. Multilevel and local time-stepping discontinuous Galerkin methods for magma dynamics. *Comput. Geosci.* **2015**, *19*, 965–978. [CrossRef]
11. Zhou, W.; Ouyang, J.; Wang, X.; Su, J.; Yang, B. Numerical simulation of viscoelastic fluid flows using a robust FVM framework on triangular grid. *J. Non-Newton. Fluid Mech.* **2016**, *236*, 18–34. [CrossRef]
12. Cheng, L.; Hu, C. An adaptive multi-moment FVM approach for incompressible flows. *J. Comput. Phys.* **2018**, *359*, 239–262.
13. Osher, S.; Sanders, R. Numerical approximations to nonlinear conservation laws with locally varying time and space grids. *Math. Comput.* **1983**, *41*, 321–336. [CrossRef]
14. Tan, Z.; Huang, Y. An Adaptive Grid Method with Local Time Stepping for One Dimensional Conservation Laws. *Nat. Sci. J. Xiangtan Univ.* **2003**, *25*, 110–116. (In Chinese)
15. Crossley, A.J.; Wright, N.G.; Whitlow, C.D. Local time stepping for modeling open channel flows. *J. Hydraul. Eng.* **2003**, *129*, 455–462. [CrossRef]
16. Crossley, A.J.; Wright, N.G. Time accurate local time stepping for the unsteady shallow water equations. *Int. J. Numer. Methods Fluids* **2005**, *48*, 775–799. [CrossRef]
17. Dazzi, S.; Maranzoni, A.; Mignosa, P. Local time stepping applied to mixed flow modelling. *J. Hydraul. Res.* **2016**, *54*, 145–157. [CrossRef]
18. Dazzi, S.; Vacondio, R.; Dal Palù, A.; Mignosa, P. A local time stepping algorithm for GPU-accelerated 2D shallow water models. *Adv. Water Resour.* **2018**, *111*, 274–288. [CrossRef]
19. Wu, D.; Yu, X.; Xu, Y. A Discontinuous Galerkin Method with Local Time Stepping for Euler Equations. *Chin. J. Comput. Phys.* **2016**, *28*, 1–9. (In Chinese)
20. Trahan, C.J.; Dawson, C. Local time-stepping in Runge–Kutta discontinuous Galerkin finite element methods applied to the shallow-water equations. *Comput. Methods Appl. Mech. Eng.* **2012**, *217–220*, 139–152. [CrossRef]
21. Hu, P.; Lei, Y.; Han, J.; Cao, Z.; Liu, H.; He, Z. Computationally efficient modeling of hydro–sediment-morphodynamic processes using a hybrid local time step/global maximum time step. *Adv. Water Resour.* **2019**, *127*, 26–38. [CrossRef]
22. Baldauf, M. Local time stepping for a mass-consistent and time split advection scheme. *R. Meteorol. Soc.* **2019**, *145*, 337–346. [CrossRef]
23. Toro, E.F. *Shock-Capturing Methods for Free-Surface Shallow Flows*; John Wiley: Hoboken, NJ, USA, 2001.
24. Thompson, J.F.; Warsi, Z.U.A.; Mastin, C.W. *Numerical Grid Generation*; North Holland: New York, NY, USA, 1985; Chapter 6; pp. 129–165.
25. Pan, C. Advanced in numerical simulation of discontinuous shallow water flows. *Adv. Sci. Technol. Water Resour.* **2010**, *30*, 77–84. (In Chinese)

26. Zhang, D.; Li, D.; Wang, X. Numerical modeling of dam-break water flow with wetting and drying change based on unstructured grids. *J. Hydroelectr. Eng.* **2008**, *27*, 98–102. (In Chinese)
27. Murillo, J.; García-Navarro, P. Wave Riemann description of friction terms in unsteady shallow flows: Application to water and mud/debris floods. *J. Comput. Phys.* **2012**, *231*, 1963–2001. [CrossRef]
28. Vacondio, R.; Dal Palù, A.; Mignosa, P. GPU-enhanced Finite Volume Shallow Water solver for fast flood simulations. *Environ. Model. Softw.* **2014**, *57*, 60–75. [CrossRef]
29. Fernández-Pato, J.; Morales-Hernández, M.; García-Navarro, P. Implicit finite volume simulation of 2D shallow water flows in flexible meshes. *Comput. Methods Appl. Mech. Eng.* **2018**, *328*, 1–25. [CrossRef]
30. Brufau, P.; Vázquez-Cendón, M.E.; García-Navarro, P. A numerical model for the flooding and drying of irregular domains. *Int. J. Numer. Methods Fluids* **2002**, *39*, 247–275. [CrossRef]
31. Lv, B.; Jin, S.; Ai, C. Well-balanced Roe-type scheme for 2D shallow water flow using unstructured grids. *Hydro-Sci. Eng.* **2010**, *2*, 39–44. (In Chinese)
32. Toro, E.F. *Riemann Solvers and Numerical Methods for Fluid Dynamics*; Springer: Berlin/Heidelberg, Germany, 2013; pp. 87–114.
33. Fennema, R.J.; Chaudhry, M.H. Explicit methods for 2-D transient free surface flows. *J. Hydraul. Eng.* **1990**, *116*, 1013–1034. [CrossRef]
34. Wang, D. *Computational Hydraulics: Theory and Application*; Science Press: Beijing, China, 2011. (In Chinese)
35. Liu, H.; Zhou, J.G.; Burrows, R. Lattice Boltzmann simulations of the transient shallow water flows. *Adv. Water Resour.* **2010**, *33*, 387–396. [CrossRef]
36. Baghlani, A. Simulation of dam-break problem by a robust flux-vector splitting approach in Cartesian grid. *Sci. Iran.* **2011**, *18*, 1061–1068. [CrossRef]
37. Hu, P.; Han, J.; Lei, Y. Coupled modeling of sediment-laden flows based on local-time-step approach. *J. Zhejiang Univ. (Eng. Sci.)* **2019**, *53*, 743–752. (In Chinese)
38. Kawahara, M.; Umetsu, T. Finite element method for moving boundary problems in river flow. *Int. J. Numer. Methods Fluids* **1986**, *6*, 365–386. [CrossRef]
39. Zhang, H.; Zhou, J.; Bi, S.; Song, L.X. Two-dimensional shallow hydrodynamic model based on adaptive structured grid. *Chin. J. Hydrodyn.* **2012**, *27*, 667–678. (In Chinese)
40. Yan, W.; Zhou, R.; Cheng, Z.; Yang, W.; Lu, H. Adaptation of fleets to the navigation discharge in the waterway between Three Gorges Project and Gezhouba Hydroproject. *J. Yangtze River Sci. Res. Inst.* **2013**, *6*, 33–37. (In Chinese)
41. Zhang, S.; Jing, Z.; Li, W.; Wang, L.; Liu, D.; Wang, T. Navigation risk assessment method based on flow conditions: A case study of the river reach between the Three Gorges Dam and the Gezhouba Dam. *Ocean Eng.* **2019**, *175*, 71–79. [CrossRef]

Article

Effect of Open Boundary Conditions and Bottom Roughness on Tidal Modeling around the West Coast of Korea

Van Thinh Nguyen * and Minjae Lee

Department of Civil and Environmental Engineering, Seoul National University, 1 Gwanak-ro, Gwanak-gu, Seoul 151-744, Korea; mjlee14@snu.ac.kr

* Correspondence: vnguyen@snu.ac.kr; Tel.: +82-2-880-7355

Received: 9 May 2020; Accepted: 11 June 2020; Published: 15 June 2020

Abstract: The aim of this study was to investigate the effect of open boundary conditions and bottom roughness on the tidal elevations around the West Coast of Korea (WCK) using an open-source computational fluids dynamics tool, the TELEMAC model. To obtain a detailed tidal forcing at open boundaries, three well-known assimilated tidal models—the Finite Element Solution (FES2014), the Oregon State University TOPEX/Poseidon Global Inverse Solution Tidal Model (TPXO9.1) and the National Astronomical Observatory of Japan (NAO.99Jb)—have been applied to interpolate the offshore tidal boundary conditions. A number of numerical simulations have been performed for different offshore open boundary conditions, as well as for various uniform and non-uniform bottom roughness coefficients. The numerical results were calibrated against observations to determine the best fit roughness values for different sub-regions within WCK. In order to find out the dependence of the tidal elevation around the WCK on the variations of open boundary forcing, a sensitivity analysis of coastal tide elevation was carried out. Consequently, it showed that the tidal elevation around the WCK was strongly affected by local characteristics, rather than by the offshore open boundary conditions. Eventually, the numerical results can provide better quantitative and qualitative tidal information around the WCK than the data obtained from assimilated tidal models.

Keywords: west coast of Korea (WCK); tidal elevation; open boundary condition; numerical modeling; TELEMAC-2D

---

1. Introduction

The Yellow Sea (YS) is located between the mainland of China and the Korean Peninsula, and its depth is about 44 m on average, with a maximum depth of 152 m. The sea bottom slope gradually rises toward the East China coast and more abruptly toward the Korean Peninsula, and the depth gradually increases from north to south. Southern YS is bounded by the East China Sea (ECS) along a line running northeastward from the mouth of the Changjiang River to the southwestern tip of Korea. ECS covers the area originating around the Taiwan Strait at about 25° N, extending northeastward to the Kyushu Island, Japan, and to the Korea Strait, and bounded by the Ryukyu Island chains on the southeastern open ocean (Figure 1). On the broad and shallow shelf under the Yellow and the East China Sea (YECS), the flow is dominated by strong semidiurnal M2 tidal currents with the superimposition of semidiurnal S2, diurnal K1 and O1 currents [1–4]. Approaching the WCK, the tidal amplitudes vary from 4 to 8 m, and the speed of the tidal current may increase to more than 1.56 m/s near the coasts.

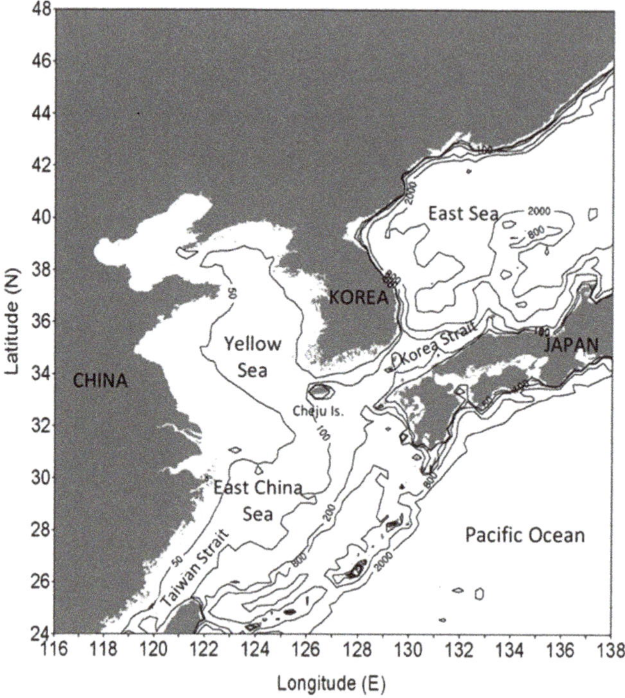

Figure 1. Geophysical configuration of the Yellow Sea and the East China Sea.

Along with strong tidal currents, the WCK is surrounded by lots of islands including extremely irregular and indented coastlines, the tidal propagation and shoaling processes on the shore in this broad flat region are very complicated and sensitive to the variation of tidal elevation.

The study on the tidal circulation in this region has been carried out by various authors, such as Le Fevre et al. [5], Kim [6], Naimie et al. [7], and Suh [8] using 2D depth-averaged finite element models on unstructured grids, or Choi [2], Tang [9], Kantha et al. [10], Kang et al. [11] and Lee et al. [12] using 3-D finite difference models on structured grids. However, there are still very few authors using data obtained from different assimilated tidal models to set up the offshore boundary conditions. Most authors have applied a constant boundary forcing condition (BFC) or tidal charts, such as Ogura's chart [1,2,13,14]) and Nishida's chart [3,15,16] or applied a very coarse observation data combined with partly constant boundary forcing conditions [11,17]. Only Le Fevre et al. [5] and Suh [8] have applied tidal charts from FES 94.1 and FES 2004 models for OBC. Using FVCOM, we recently applied a tidal chart from NAO.99Jb. In principle, the tidal information around the WCK can be obtained from assimilated global and regional tidal models [18].

In principle, the tidal information around the WCK can be obtained from assimilated global and regional tidal models. The assimilation model, which is a combination of a dynamical model and observed data, was actively advanced in the research group of the modern global tide model, and pioneered by [19]. By the assimilation model, the problem of the empirical method with regard to the resolution can be compensated by a hydrodynamic model, and any inadequate potential of the hydrodynamic model can be recovered by observed data [20]. Recently, using the unprecedented sea surface height data of satellite altimeter TOPEX/POSEIDON (T/P), several research groups have developed highly accurate assimilation models especially in an open ocean ([5,20–24]). T/P was launched on 10 August 1992, which measures sea level relative to the ellipsoid with an overall accuracy of approximately 5 cm [25]. Due to the high accuracy of T/P altimeter data for those assimilation models, it can be used to extract relatively accurate tidal boundary conditions around the marginal sea within YECS.

Among the global assimilation models, FES2014 (https://www.aviso.altimetry.fr/en/data/products/auxiliary-products/global-tide-fes.html) and TPXO9.1 (http://volkov.oce.orst.edu/tides/tpxo9_atlas.html) atlases are open to the public and can be freely downloaded. The FES2014 atlas was produced by Legos and CLS Space Oceanography Division, and distributed by Aviso, with support from CNES (https://datastore.cls.fr/catalogues/fes2014-tide-model/). It is distributed as a tidal chart on grid resolution of 1/16° based on the dynamical finite element tide model CEFMO, and data assimilation code CADOR. The finite element grid of the model CEFMO has about 2,900,000 computational nodes and covered 34 tidal constituents. In addition to above two global tide models, the regional tide model NAO.99Jb (https://www.miz.nao.ac.jp/staffs/nao99/index_En.html) was developed by Matsumoto et al. (2000) for the region around Japan with a resolution of 1/12°, covering from 110 to 165° E and from 20–65° N so that YECS was entirely included in this region. This model was nested with the global model NAO.99b to provide open boundary conditions to the regional model and the assimilation was performed with both T/P data and 219 coastal tidal gauges obtained around Japan and Korea. Most T/P-derived altimetric tides can be inaccurate in shallow coastal oceans because the resolution enforced on data analysis by the spatial resolution of T/P is simply inadequate near ocean margins.

Therefore, we first validated the tidal information of tidal constituents K_1, O_1, M_2 and S_2 provided by three assimilated tidal models; FES2014, NAO99Jb and TPXO9.1 against the observation at 21 tidal gauges along WCK. Figure 2 shows the comparison of amplitude (left) and phase (right) obtained from the assimilated tidal models with observation data obtained from gauging stations along the WCK. It shows that while the phases are better agreement with observation, the amplitudes obtained from the assimilated tidal models still poorly agreed with the observed data. More detailed quantitative comparisons between the amplitude and phase obtained from the assimilated tidal models and observation data are shown in Table A1 (in Appendix A). Wherein it shows NAO99Jb provides a better agreement of the amplitude for the constituents M_2 and S_2, while FES2014 provides a little better fit of the amplitude for K_1 and O_1. It also shows the constituent M2 has the largest amplitude in comparison with other tidal constituents; about 5–8 times, 2.5–2.8 times, and 6.6–11.8 times larger than K1, S2 and O1 amplitudes, respectively.

In addition, the resolutions (1/6°, 1/12° and 1/16°) of three tidal models are not fine enough along the coastlines, so they cannot provide enough detailed quantitative and qualitative tidal information along the WCK. Therefore, in order to obtain more detail on tidal information along the WCK, an accuracy numerical simulation with higher grid resolutions is needed.

This study is to introduce several numerical investigations using higher grid resolutions and applying different open boundary conditions extracted from three well-known assimilated tidal models FES2014, TPXO9.1 and NAO.99Jb then calibrated with different uniform and non-uniform bottom roughness values for different sub-regions within WCK to investigate the effect of open boundary condition and bottom roughness on the tidal elevations from major constituents (M2, K1, S2 and O1) around the West Coast of Korea.

Based on the form number (F), the ratio of the amplitudes of K_1, O_1, M_2, *and* S_2 suggested by [26] is shown in Equation (1).

$$F = \frac{K_1 + O_1}{M_2 + S_2} \quad (1)$$

whereby the mixed tide is classified following the value of the form number F; if the value of F less than 0.25 or larger than 1.25, then it is classified as semidiurnal or diurnal, respectively. Otherwise, if the value of F is located between 0.25 and 1.25, it is considered mixed.

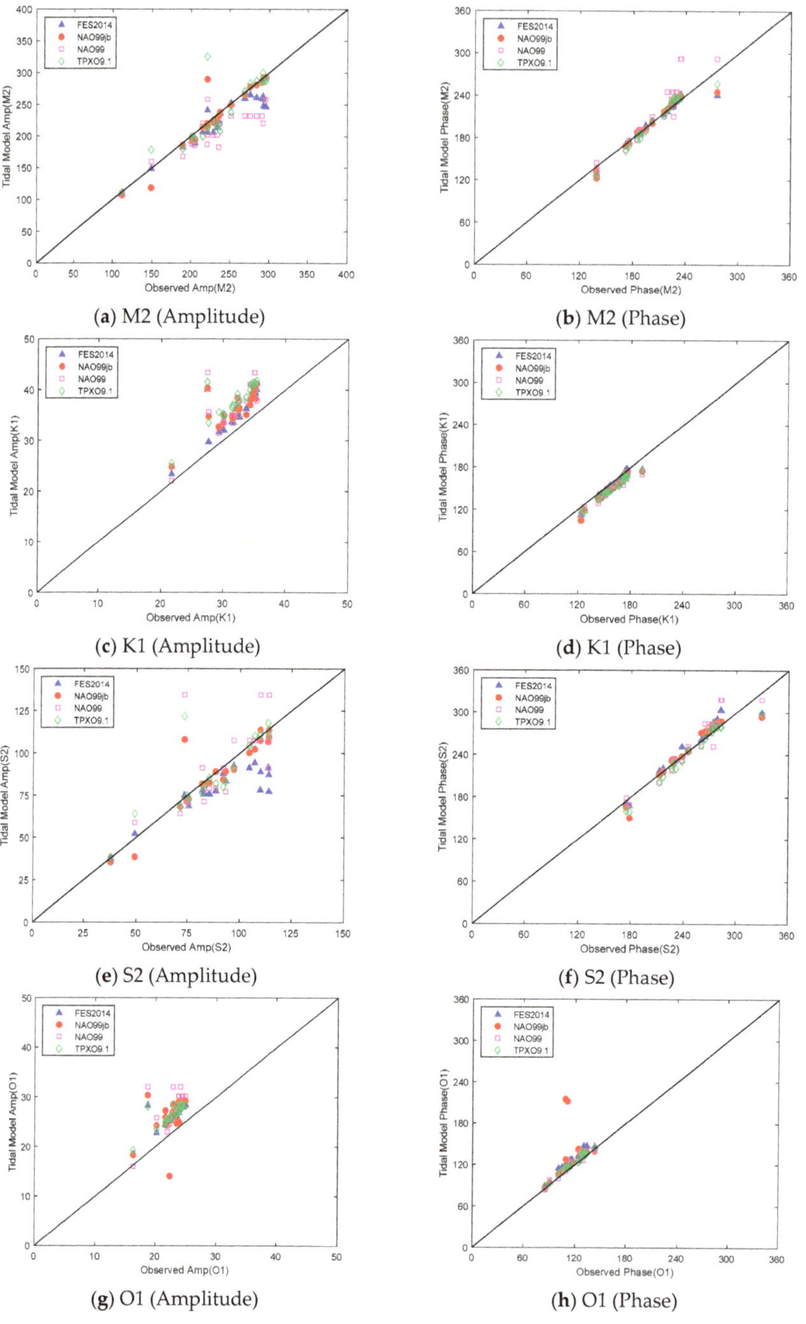

(**a**) M2 (Amplitude) (**b**) M2 (Phase)

(**c**) K1 (Amplitude) (**d**) K1 (Phase)

(**e**) S2 (Amplitude) (**f**) S2 (Phase)

(**g**) O1 (Amplitude) (**h**) O1 (Phase)

Figure 2. Validation of amplitudes and phases obtained from assimilated tidal models (FES2014, NAO99Jb, NAO99 and TPXO9) against observed data at tidal gauges; (**a**): M2's amplitude, (**b**): M2's phase, (**c**): K1's amplitude, (**d**): K1's phase, (**e**): S2's amplitude S2, (**f**): S2's phase, (**g**): O1's amplitude, (**h**): O1's phase.

Figure 3 shows the values of the form number along the WCK are below 0.25 according to the data obtained from observation at 21 tidal gauges, as well as from three assimilated tidal models, FES2014, TPXO9.1 and NAO.99Jb; therefore, the mixed tide prevails semidiurnal in the WCK.

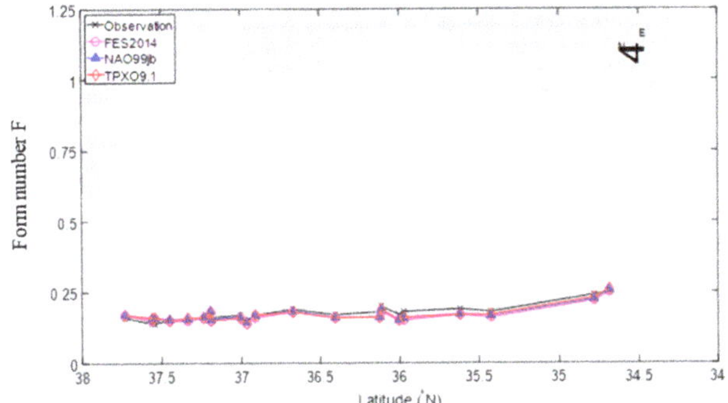

Figure 3. The value of the form number (*F*) in the WCK. Grey color in the background shows the WCK, from left to right corresponding to north to south.

2. Numerical Modeling

As mentioned above, since the study region was quite shallow, the numerical simulation was based on an open-source software, the TELEMAC-2D (http://www.opentelemac.org) model. It is well-known software with an abundant history of development and application over many years in fluvial and maritime hydraulics [27,28]. The TELEMAC-2D is capable of capturing the complicated bathymetry and coastlines of the WCK surrounded by lots of islands including extremely irregular and indented coastlines by using unstructured grids.

2.1. Basic Equations

The TELEMAC-2D has solved the depth-averaged Navier-Stokes equations with two-equation turbulence closure models (k-ε) using the finite element method, as follows:

The continuity equation

$$\frac{\partial h}{\partial t}+\frac{\partial (hU)}{\partial x}+\frac{\partial (hV)}{\partial y}=0 \tag{2}$$

and momentum equations

$$\frac{\partial (hU)}{\partial t}+\frac{\partial (hUU)}{\partial x}+\frac{\partial (hUV)}{\partial y}=-h\cdot g\frac{\partial Z}{\partial x}+h\cdot F_x+\mathrm{div}\left(h\cdot \nu_e\cdot \vec{\nabla}(U)\right) \tag{3}$$

$$\frac{\partial (hV)}{\partial t}+\frac{\partial (hUV)}{\partial x}+\frac{\partial (hVV)}{\partial y}=-h\cdot g\frac{\partial Z}{\partial y}+h\cdot F_y+\mathrm{div}\left(h\cdot \nu_e\cdot \vec{\nabla}(V)\right) \tag{4}$$

$$\frac{\partial k}{\partial t}+U\frac{\partial k}{\partial x}+V\frac{\partial k}{\partial y}=\frac{1}{h}\,\mathrm{div}\left(h\cdot \frac{\nu_t}{\sigma_k}\vec{\nabla}(k)\right)+P-\varepsilon+P_{kv} \tag{5}$$

$$\frac{\partial \varepsilon}{\partial t}+U\frac{\partial \varepsilon}{\partial x}+V\frac{\partial \varepsilon}{\partial y}=\frac{1}{h}\,\mathrm{div}\left(h\cdot \frac{\nu_t}{\sigma_k}\vec{\nabla}(\varepsilon)\right)+\frac{\varepsilon}{k}(c_{1\varepsilon}P-c_{2\varepsilon}\varepsilon)+P_{\varepsilon v} \tag{6}$$

where h is the water depth; U and V are the depth averaged velocity components;

$U=\frac{1}{h}\int_{Z_f}^{Z}udz$ and $V=\frac{1}{h}\int_{Z_f}^{Z}vdz$

Z and Z_f are the free surface and bottom elevations, respectively;

F_x and F_y are body forces in x and y directions including Coriolis (F_x^c, F_y^c), bottom friction F_x^f, F_y^f), and surface wind F_x^w, F_y^w) forces determined as $F_x=F_x^c+F_x^f+F_x^w$ and $F_y=F_y^c+F_y^f+F_y^w$

Coriolis force; $F_x^c = 2\omega\sin(\lambda)v$ and $F_y^c = -2\omega\sin(\lambda)u$; $\omega = 7.292\times10^{-5}$ is angular and λ is latitude;

Bottom friction force; $F_x^f = -\frac{1}{2h}C_f U\sqrt{U^2+V^2}$ and $F_y^f = -\frac{1}{2h}C_f V\sqrt{U^2+V^2}$, where C_f is the bottom friction coefficient, $C_f = \frac{gn^2}{h^{1/3}}$, n is Manning's coefficient;

Wind forcing on the free surface is neglected; i.e., $(F_x^w, F_y^w) = 0$.

And v_e is the effective viscosity, the summation of the molecular viscosity v and the turbulent viscosity v_t $v_e = v + v_t$; where

$$v_t = c_\mu \frac{k^2}{\varepsilon}. \tag{7}$$

The depth averaged kinetic energy k and its dissipation rate ε are:

$$k = \frac{1}{h}\int_{Z_f}^{Z}\frac{1}{2}\overline{u_i'u_i'}dz \text{ and } \varepsilon = \frac{1}{h}\int_{Z_f}^{Z} v\overline{\frac{\partial u_i'}{\partial x_j}\frac{\partial u_i'}{\partial x_j}}dz$$

where u_i' is the fluctuating velocity and the overbar represents an average over time; i.e., $u_i = U_i + u_i'$; U_i (=U or V) is an average over time of velocity components; and P is the production term:

$$P = v_t\left(\frac{\partial U_i}{\partial x_j} + \frac{\partial U_j}{\partial x_i}\right)\frac{\partial U_i}{\partial x_j} \tag{8}$$

P_{kv} and $P_{\varepsilon v}$ are vertical shear terms: $P_{kv} = c_k\frac{u_*^3}{h}$; $P_{\varepsilon v} = c_\varepsilon\frac{u_*^4}{h^2}$ where $c_k = \frac{1}{\sqrt{c_f}}$ and $c_\varepsilon = 3.6\frac{c_{2\varepsilon}\sqrt{c_\mu}}{c_f{}^{3/4}}$

u_* is the friction velocity: $u_* = \sqrt{\frac{C_f}{2}(U^2+V^2)}$

The empirical constants in Equations (5) and (6) $C_\mu = 0.09$, $C_{1\varepsilon} = 1.44$, $C_{2\varepsilon} = 1.92$, $\sigma_k = 1.0$ *and* $\sigma_\varepsilon = 1.3$ of the k-ε model are obtained from classical test cases.

2.2. *Boundary Conditions*

2.2.1. At the Solid Boundary

On the solid boundary, such as coastline and bottom, the no-slip condition is applied; i.e., velocity $\vec{U} = (U,\ V) = (0,\ 0)$.

In this region, particularly near the shallow coastline, the effect of bottom friction becomes prominent, the bottom friction is calculated by:

$$\tau_b = \left(\tau_{bx},\ \tau_{by}\right) = -\frac{1}{2}\rho C_f\sqrt{U^2+V^2}(U,\ V) \tag{9}$$

The value of the coefficient C_f is obtained from the calibration procedure, which is shown in Section 3.1 below.

The boundary condition for turbulence kinetic energy and its dissipation are defined as

$$k_\delta = \frac{u_*^2}{\sqrt{C_\mu}} + \frac{C_{2\varepsilon}}{C_\varepsilon C_f}u_*^2 \text{ and } \varepsilon_\delta = \frac{u_*^3}{\kappa\delta} + \frac{1}{\sqrt{C_f}}\frac{u_*^3}{h} \tag{10}$$

where δ is a distance from the wall, $\delta = 0.1$ of a local mesh size; u_* is friction velocity of the wall.

2.2.2. At the Free Surface

We assume the wind force on the free surface is neglected as mentioned above, i.e., $F_x^w = F_y^w = 0$.

2.2.3. At the Open Boundary

Tidal harmonic parameters obtained from each assimilated tidal models FES2014, NAO99Jb and TPXO9.1 were combined into the open boundary conditions, as follows: $H = \sum_i H_i$

$$H_i(M,\ t) = A_{F_i}(M)\cos\left(2\pi\frac{t}{T} - \varphi_{F_i}(M)\right) \tag{11}$$

where H_i is the water elevation, A_{F_i} is amplitude; T is period and φ_{F_i} is phase of each tidal constituent.

In this study, four major tidal constituents M2, K1, S2 and O1, were considered. The harmonic constants of tidal constituents for open boundary conditions were obtained from three assimilated tidal models; FES2014, TPXO9.1 and NAO.99Jb.

Turbulence kinetic energy and its dissipation at the open boundary were set as: $k = C_{2\varepsilon}\frac{P_{kv}^2}{P_{\varepsilon v}}$ and $\varepsilon = P_{kv}$.

2.3. *Study Region*

The study area covered the entire YS and most of the ECS in spherical coordinate. The bathymetry data of 30 arc-seconds resolution was downloaded from the General Bathymetric Chart of the Ocean (GEBCO2014: https://www.ngdc.noaa.gov/mgg/ibcm/ibcmdvc.html) as shown in Figure 4. The shorelines with an accuracy of about 250 m were downloaded from the NOAA shoreline website (https://shoreline.noaa.gov). This region included the flat and broad continental shelf of YECS, bounded by Taiwan Strait (A), a line of shelf-break determined by 200 m isobaths (B), and Korea Strait (C), as shown in Figure 5. The open boundary was expanded to the shelf edge of YECS which was advantageous to use offshore co-oscillating tidal conditions as a boundary forcing in order to minimize disturbances by the geographical configuration and nonlinear tidal response near-shore. Open boundaries adjacent to the offshore also enabled the numerical results around WCK to minimize the influence of unexpected numerical instabilities at open boundaries.

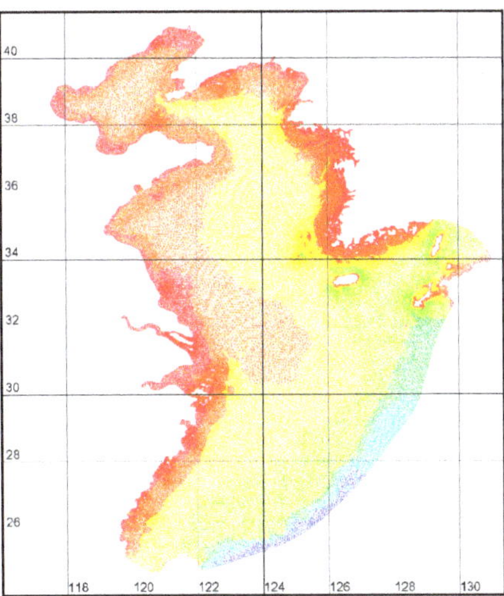

Figure 4. Bathymetry of the YECS Region.

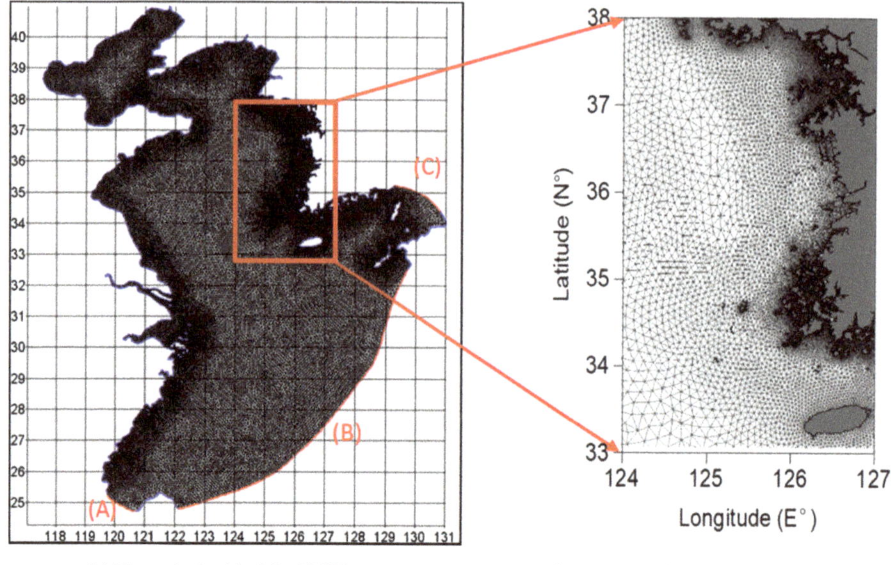

(**a**) Numerical grid of the YECS (**b**) Numerical grid of the WCK

Figure 5. Horizontal unstructured grid of the YECS (**a**) and an enlargement of WCK (**b**).

In this study, the numerical simulation applied a hybrid horizontal resolution of an unstructured grid using a free pre-processing Blue Kenue software tool (https://nrc.canada.ca/en/research-development/products-services/software-applications/blue-kenuetm-software-tool-hydraulic-modellers). Herein the unstructured triangular grid had a horizontal resolution varying from 0.2 to 1 km in the WCK, 3–5 km in the East coast of China, and 9–12 km in the interior of YS and ECS. After running a number of simulations to check for grid independence, the final grid of 203,927 grid cells was used for the simulation in this study. Bathymetry was interpolated to each nodal point of the horizontal grid using Blue Kenue software as well (Figure 5).

A total of 21 tidal gauges were collected around the WCK to provide the observed tidal elevation data of semidiurnal M2 and S2 and diurnal K1 and O1 constituents in YECS. As shown in Figure 6, the WCK can be divided into three sub-regions based on the common characteristics of their bathymetry and coastlines—the first region was Kyunggi Bay near Incheon (I), the second region was around the estuary of the Geum River near Gunsan (G), and the third region was around the south-western tip near Mokpo (M). The numbers of tidal gauges located in three regions Mokpo, Gunsan and Incheon were 2, 8 and 11, respectively. The most recent observation data in 2017 downloaded from Korea Hydrographic and Oceanographic Agency (http://www.khoa.go.kr/koofs/kor/observation/obs_real.do) have been selected for the calibration and validation of numerical models.

Figure 6. Tidal gauge locations in the WCK selected for the evaluation of the modeling; three sub-regions: Mokpo (M), Gunsan (G) and Incheon (I) Regions.

3. Numerical Results and Discussion

The numerical simulations were carried out on our high-performance computing cluster (http://cfdlabsnu.com) for the real-time simulation of 60 days with time step of 20 s. The numerical results obtained from the last 30 days simulation were used for the following analyses.

3.1. *Response of the Tide around WCK to the Bottom Roughness*

3.1.1. Applying Uniform Roughness Coefficient

Previous tidal calculations in the YECS region were mostly using the quadratic bottom friction law, in which the typical value of roughness coefficient C_f was chosen as a uniform value of 0.0025, as shown in [2,16,29,30]. We again have verified the numerical model with various uniform roughness coefficients that ranged from 0.0015 to 0.03 for different offshore open boundary conditions (OBCs) interpolated from three assimilated tidal models (FES2014, TPXO9.1 and NAO.99Jb).

The WCK was divided into three different sub-regions—Mokpo (M), Gunsan (G), and Incheon (I)—as shown in Figure 7, for applying different bottom friction coefficients. The borderline of this sub-regions was set based on a water depth contour of around 50 m. Different values of bottom friction coefficient were applied with regards to the geographic characteristics.

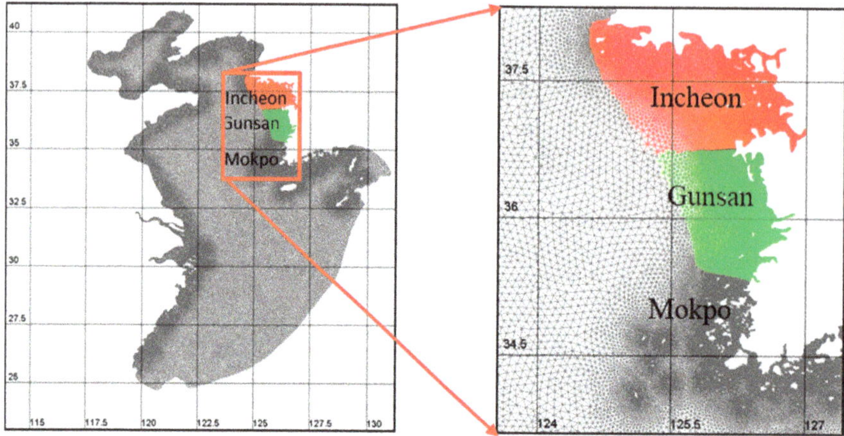

Figure 7. Sub-dividing regions in WCK based on roughness coefficients.

As shown in Tables A2–A4 (in Appendix A), based on the evaluation of root mean square deviation (RMSE) of tidal amplitudes between the observation and the numerical results obtained from applying various uniform roughness coefficients, the results show that the numerical results were very distinct from each other. For the tidal constituent M2, the bottom roughness coefficient $C_f = 0.0023$ provides the best result obtained from FES2014 and TPXO9.1, while the numerical results obtained from NAO99Jb were not significantly different between $C_f = 0.0023$ and 0.0025. For the tidal constituent K1, the bottom roughness coefficient $C_f = 0.002$ provides the best result obtained from NAO99Jb, while the numerical results obtained from FES2014 were not significantly different between $C_f = 0.0015$ and 0.002; the numerical results obtained from TPXO9.1 were not significantly different between $C_f = 0.002$, 0.0023 and 0.0025. For the tidal constituent S2, the bottom roughness coefficient $C_f = 0.0015$ provides the best result obtained from all three assimilated tidal models. For the tidal constituent O1, the bottom roughness coefficient $C_f = 0.0015$ provides the best result obtained from FES2014 and TPXO9.1, while the numerical results obtained from NAO99Jb were not significantly different between $C_f = 0.0015$ and 0.002. In addition, based on the RMSE values in Tables A2–A4, the results show that for M2 the most appropriate roughness values for Mokpo, Gunsan and Incheon regions were 0.0025, 0.003 and 0.0023, respectively. Whereas, for S2, the most appropriate roughness values for Mokpo, Gunsan and Incheon regions were 0.002, 0.0023 and 0.0015, respectively. For O1, the most appropriate roughness value for Mokpo, Gunsan and Incheon regions was 0.0015 and for K1 the most appropriate roughness values with FES2014 and NAO99Jb for Mokpo, Gunsan and Incheon regions were 0.0015, 0.002 and 0.002, respectively. However, with TPXO9.1 the values for Mokpo, Gunsan and Incheon regions were 0.002, 0.0023 and 0.002, respectively.

3.1.2. Applying Non-Uniform Roughness Coefficient

As shown above, applying a uniform bottom roughness coefficient for the entire WCK was inappropriate. According to the bathymetry data, the bottom slope was relatively milder in the Incheon region than in Mokpo and Gunsan regions. In addition, there were lots of islands around the Mokpo region rather than the Incheon region. We can expect that more energy dissipation occurred in the Mokpo and Gunsan regions than in the Incheon region. The larger values of the bottom friction coefficient were applied to Mokpo and Gunsan regions than the Incheon region. In addition, the roughness coefficient values for the Gunsan region were set to be the largest because its bed form was more variable than the Mokpo region.

Table A5 shows the list of simulation cases applying non-uniform bottom roughness coefficients to different sub-regions Mokpo, Gunsan, Incheon and other regions within WCK. As shown in Tables A6–A9, once the non-uniform bottom roughness coefficients were applied, the mean RMSE values for M2 from the varied ranged of 14–36 cm down to a value of about 11 cm with OBCs obtained from FES2014 and TPXO9.1. These values varied from 16–41 cm down to 12–13 cm with OBCs obtained from NAO99Jb while, there was no significant improvement for K1, S2 and O1. In the WCK, M2 was the dominant constituent, followed by S2 and K1, as mentioned by Teague et al. (1998). In addition, the semidiurnal tides M2 and S2, as well as the diurnal tides K1 and O1, had similar tendencies responding to the local effects and OBCs. Therefore, M2 was presented for the semidiurnal tide, and K1 presented for diurnal tides in most following numerical results. The comparisons of M2 and K1 amplitudes between the observations and simulation results with uniform and non-uniform bottom roughness coefficients are shown in Figures 8 and 9. It clearly shows the simulation results obtained from non-uniform bottom roughness coefficients have significantly improved the results for M2 and K1.

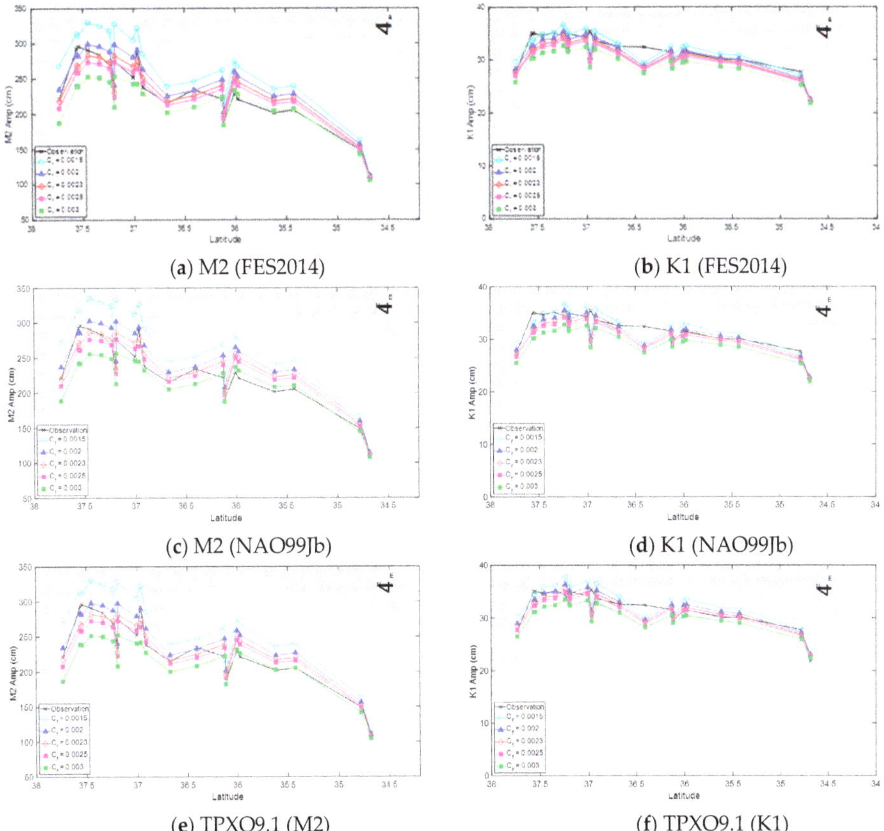

(**a**) M2 (FES2014) (**b**) K1 (FES2014)
(**c**) M2 (NAO99Jb) (**d**) K1 (NAO99Jb)
(**e**) TPXO9.1 (M2) (**f**) TPXO9.1 (K1)

Figure 8. The numerical results of tidal amplitudes M2 and K1 obtained from different uniform bottom roughness; (**a**), (**c**) and (**e**): M2's amplitude; (**b**), (**d**) and (**f**): K1's amplitude.

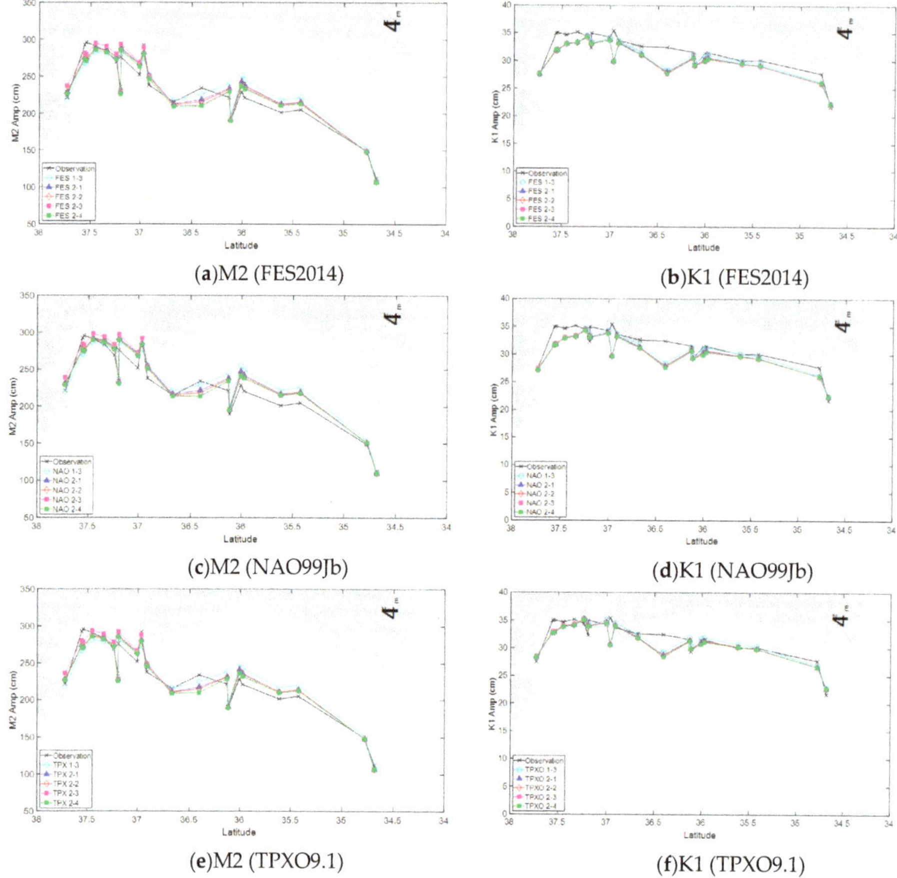

(a)M2 (FES2014) (b)K1 (FES2014)

(c)M2 (NAO99Jb) (d)K1 (NAO99Jb)

(e)M2 (TPXO9.1) (f)K1 (TPXO9.1)

Figure 9. The numerical results of tidal amplitudes M2 and K1 obtained from non-uniform bottom roughness; (**a**), (**c**) and (**e**): M2's amplitude; (**b**), (**d**) and (**f**): K1's amplitude.

The bottom friction coefficient is a physical parameter characteristic of bottom materials and bathymetries. Therefore, it cannot vary following each tidal constituent. From Tables A5–A9, an acceptable selection from all test cases, which can deliver reasonable results for all tidal constituents (M2, K1, S2 and O1), was the case name "FES/NAO/TPX 2-4" whereby the bottom friction coefficient C_f took a value of 0.0025 for the Mokpo region, 0.0035 for the Gunsan region, and 0.002 for Incheon and other regions. These values are in the range suggested by [31,32] applying also for four tidal constituents (M2, K1, S2 and O1).

3.2. Response of the Tide around the WCK to the Open Boundary

As mentioned in Section 2.2 above, the modeling region has been bounded by three different open boundaries; A, B and C, as shown in Figure 3. Averaged tidal amplitudes at each boundary A, B and C obtained from three tidal models are shown in Table A10, in which the nodal boundary amplitudes are averaged with weighting by the length of each segment of boundary cells. The tidal amplitudes along the boundary A were significantly larger than the values along the boundaries B and C. Particularly, the tidal amplitude of M2 along the boundary A were larger than the values along the boundaries B and C; about 3.3 times and 9.3 times, respectively. The tidal amplitudes at the boundary C were the smallest because when the tides propagated from the deep region of the East Sea (its depth was about 2000 m) through boundary C via the shallow region at the Korea Strait (its depth was less than 80 m), the tidal amplitudes were significantly damped due to shoaling process on this shallow shelf region.

Table A11 shows the difference of the interpolated tidal amplitudes at three open boundaries (A, B and C) obtained from two global tidal models (FES2014 and TPXO9.1) in comparison with those obtained from the regional tidal model (NAO.99Jb). It shows that the amplitudes of M2, K1, S2 and O1 obtained from three tidal models were not substantially different at the open boundaries A, B and C.

Nevertheless, it poses a question whether the open boundary condition at A, B or C will play a substantial role on the tide around the WCK. To identify such influence, we have evaluated the response of coastal tide to open boundary forcing in the WCK, using the sensitivity analysis in the following sections.

3.2.1. Sensitivity Analysis

Response of the Tide around the WCK to Individual Boundary Forcing

The first numerical test was performed to find out how the co-oscillation of each boundary effected the tidal elevation along the WCK using uniform forcing at open boundaries A, B and C.

Assuming uniform boundary tide constituent for M2 or K1, an amplitude of 100 cm was forced on each boundary A, B, and C, individually. The resultant amplitudes for all gauge locations obtained from each open boundary are shown in Figure 10.

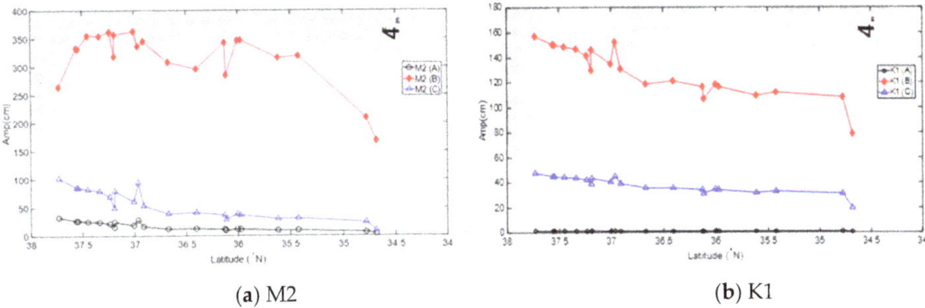

(**a**) M2 (**b**) K1

Figure 10. Computed amplitudes at gauge locations with respect to the assumed forcing with an amplitude of 100 cm at each open boundaries A, B and C; (**a**): M2's amplitude; (**b**): M2's amplitude.

It shows that only the tidal wave from the open boundary B produced much higher coastal amplitude than other incident boundary waves from open boundaries A and C. Whereby the tidal amplitudes of M2 and K1 approached 350 and 160 cm in the Kyunggi Bay within area I, respectively; whereas the force propagated from the boundaries A and C was damped on the way to the WCK before it approached the coast. Defant [33] also noticed that the tidal phenomenon in the entire ECS was almost exclusively conditioned by those water-masses which penetrated through the canals between the Ryukyu Islands. Figure 10a, shows the trend of M2 along the WCK obtained from the forcing at only open boundary B was similar to the results shown in Figure 9a,c,e, in which the real forcing at all three open boundaries A, B and C was taken into account. Therefore, the accuracy of boundary condition on B played a major role in the accuracy of the whole modeling results even when the real tidal condition was fully specified on three boundaries. As a result, in the following section, we have estimated the sensitivity of open boundary forcing at the boundary B on the tide around the WCK, instead of taking all three open boundaries into consideration.

Response of the Tide around the WCK to the Tidal Amplitude at Open Boundary

For the quantitative assessment of the influence of tidal amplitudes at the open boundary on the coastal tide elevation, a mean increase of amplitude at each gauge location with respect to the variation of open boundary amplitude was calculated by

$$D_{a,i} = \frac{r_i(a_2) - r_i(a_1)}{a_2 - a_1} \tag{12}$$

where a_1 and a_2 are different boundary tidal amplitudes in comparison, and $r_i(a_j)_{j=1,2}$ is a response of coastal amplitude at the *i-th* gauge location.

An increase of tidal amplitudes of constituents M2 and K1 was averaged over gauge locations within a specified region (I, G or M). As shown in Table A12, once the boundary amplitude of M2 constituent increased from 20 to 40 cm, 40 to 60 cm, 60 to 80 cm, and 80 to 100 cm, the mean coastal amplitude increased 3.25, 2.62, 2.01 and 1.91 cm per 1 cm increase in the open boundary amplitude, respectively. It shows that the higher the increment of the boundary amplitude, the lower the mean increase of the coastal amplitude per 1 cm. Meanwhile, the mean increase of the coastal amplitude of K1 was about 1.0 cm (ranged from 0.91 to 1.10), once the boundary amplitude of K1 constituent increased from 20 to 40 cm, 40 to 60 cm, 60 to 80 cm, and 80 to 100 cm.

Figure 11 shows the coastal M2 and K1 amplitudes at gauge locations corresponding to forcing with increasing amplitudes of 20, 40, 60, 80 and 100 cm at open boundary B. It shows that as the amplitudes increased at open boundary B, the M2 amplitude significantly increased once it entered the Mokpo region (M), and held this tendency until it reached to the estuary of Geum River, then it dropped until leaving the Gunsan region (G). Thereafter, it started to increase again when entering the Incheon regions (I) reaching the maximum value within Kyunggi Bay, and then significantly decreased. It means that the M2 amplitude was very sensitive within Kyunggi Bay. While K1 was linearly increased a little when it entered the WCK.

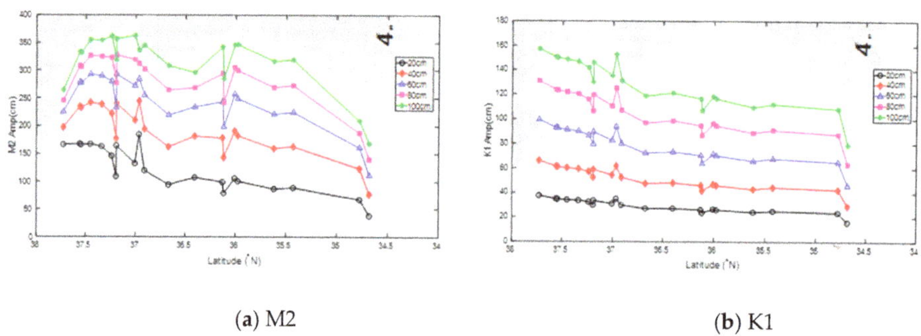

(**a**) M2 (**b**) K1

Figure 11. M2 and K1 amplitudes at gauge locations with respect to the forcing with the amplitudes of 20, 40, 60, 80 and 100 cm at the open boundary B; (**a**): M2's amplitude; (**b**): M2's amplitude.

In addition, Figure 11 also shows that once the tidal amplitude of M2 at boundary B was about 20 cm, the resultant amplitudes propagated within Kyunggi Bay had little different tendencies in comparison with those obtained from the amplitude larger than 20 cm, since it was not remarkably damped.

Table A12 also shows that regions I and G were most sensitive to the boundary amplitude for M2 simulation, once its amplitude was less than 60 cm. Once the amplitude of M2 increased more than 60 cm at open boundary B, it caused less impact on the increase of M2 amplitude along the WCK, since the value of D_a did not significantly increase. Meanwhile, an increase of K1 amplitude at the boundary B with an increment of 20 cm caused less significant change on mean D_a within all three regions of WCK.

Due to the nonlinearity in the advection terms of the governing equation, and the tidal dissipation by the nonlinear bottom shear stress on the shallow water, the increasing coastal amplitude was found to not be proportional to the increase of offshore tidal amplitude. For a more detailed analysis of the response of WCK to the tidal amplitude at open boundary B, the nonlinear response of tidal amplitude within WCK was evaluated by the following sensitivity estimation:

$$R_{a,i} = \frac{r_i(a_2)/r_i(a_1)}{a_2/a_1} \tag{13}$$

As shown in Figure 12, until the tide approached the tip of the region M, in which $R_{a,i}$ was close to 1, the M2 amplitude increased proportionally to the boundary amplitude. However, when it propagated along the coastline, the increase of amplitude started to depress. As it was pointed out, a substantial depression of amplitudes occurred at some locations in the region M. The sensitivity $R_{a,i}$ dropped mildly during tide propagation through the region G, then it significantly dropped in the region I. Being different from the M2 constituent, the coastal K1 amplitude was less sensitive in the entire WCK after the tip of region M.

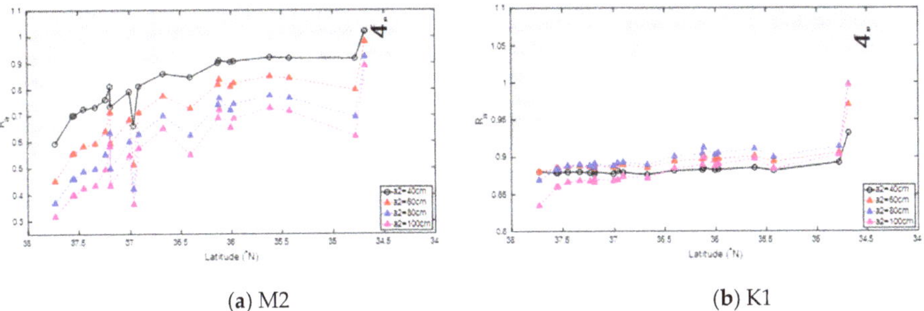

(**a**) M2 (**b**) K1

Figure 12. Value R_a at tide station locations with respect to the increase of the amplitude of M2 (**a**) and K1 (**b**) at the open boundary B; (**a**): R_a of M2's amplitude; (**b**): R_a of K1's amplitude.

From the sensitivity analysis in this section, we can estimate the response of coastal amplitude in the WCK to the difference in open boundary amplitudes of the three established boundary conditions. As shown in Table A10, the averaged amplitude obtained from three different offshore models on the boundary B was in the range from 55.5 to 56.9 cm for M2, and from 21.4 to 21.6 cm for K1. As mentioned in the previous section, the variation of mean coastal amplitude on total gauges was approximated from 1.91 to 3.25 cm for M2, and from 0.91 to 1.10 cm (about 1 cm) for K1 per 1 cm of boundary amplitude variation shown in Table A12, respectively. In Table A11, the RMSE difference in tidal amplitude between the interpolated boundary conditions by two global models (FES2014 and TPXO9.1) and the regional model (NAO99Jb) was found to be 1.85 and 2.78 cm for M2; 0.41 and 0.42 cm for K1; respectively at the open boundary B.

3.3. Comparison between the Numerical Results and Observations

As mentioned above, many authors such as [1–3] have carried out the tidal simulations in the Yellow Sea and East China Sea (YECS) region using only four dominant tidal constituents; M2, K1, S2 and O1. Particularly, [4] analyzed 13 significant tidal constituents (M2, K1, S2, O1, MM, P1, MU2, N2, MKS2, L2, K2, M4 and MS4) to conclude that M2 was the dominant constituent, followed by S2 and K1 in this region. In this study, we continued to conduct the tidal modeling for this region, taking into account four dominant tidal constituents.

Figure 13 shows a comparison of amplitudes between the observations and numerical results of M2 and K1 at gauge stations. It shows a great improvement in comparison to Figure 2, and the numerical results implying different OBCs obtained from three assimilated tidal models (FES2014, TPXO9.1 and NAO.99Jb) were similar.

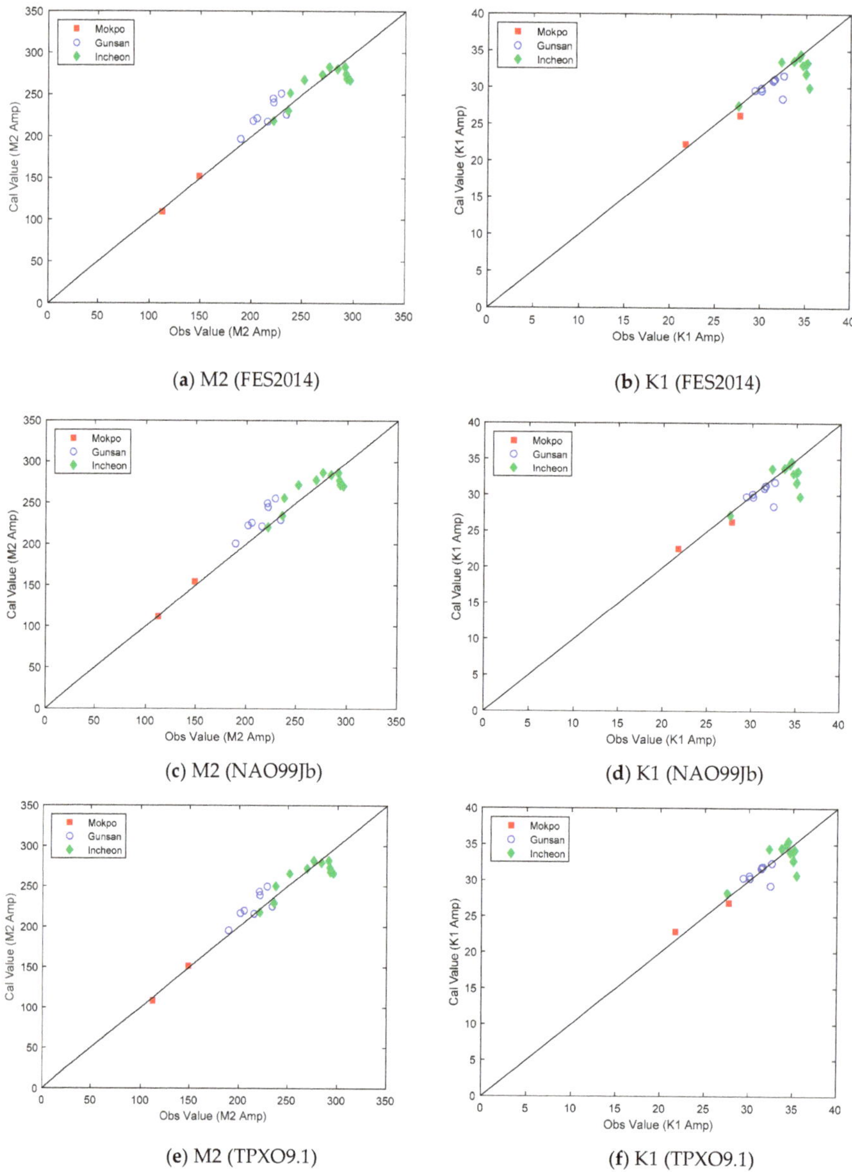

Figure 13. Comparison of the amplitudes between the observation and numerical simulation at tidal gauge stations (refer to Figure 6); (**a**,**c**,**e**): M2's amplitude; (**b**,**d**,**f**): K1's amplitude.

A question has arisen regarding the difference between the real water level observed around the WCK in comparison with the water level obtained from the numerical results contributed to by only four dominant tidal constituents. Figure 14 below shows a comparison of water level between numerical results and observations at typical gauge stations in three different regions (Incheon, Gunsan and Mokpo). In detail, Table A14 shows a comparison of water levels between the observations and numerical results at 21 gauge stations. Overall, it shows that the water levels obtained from the simulation were overestimated for the neap tide, and underestimated for spring tide. Most stations show the difference in water level was below 20%; at some specific stations located in the Incheon Region, the difference was up to 30%. Nevertheless, as mentioned by the study of Teague et al. [4], after analysis on the contribution of 13 significant tidal constituents on the water level in this region, they noticed that the tides were found to account for at least 85% of the sea surface height variance in the Yellow Sea. Therefore, the water level obtained from the numerical results was reasonable, since its differences compared to the observations can be partly compensated with the amount of 15% mentioned in [4].

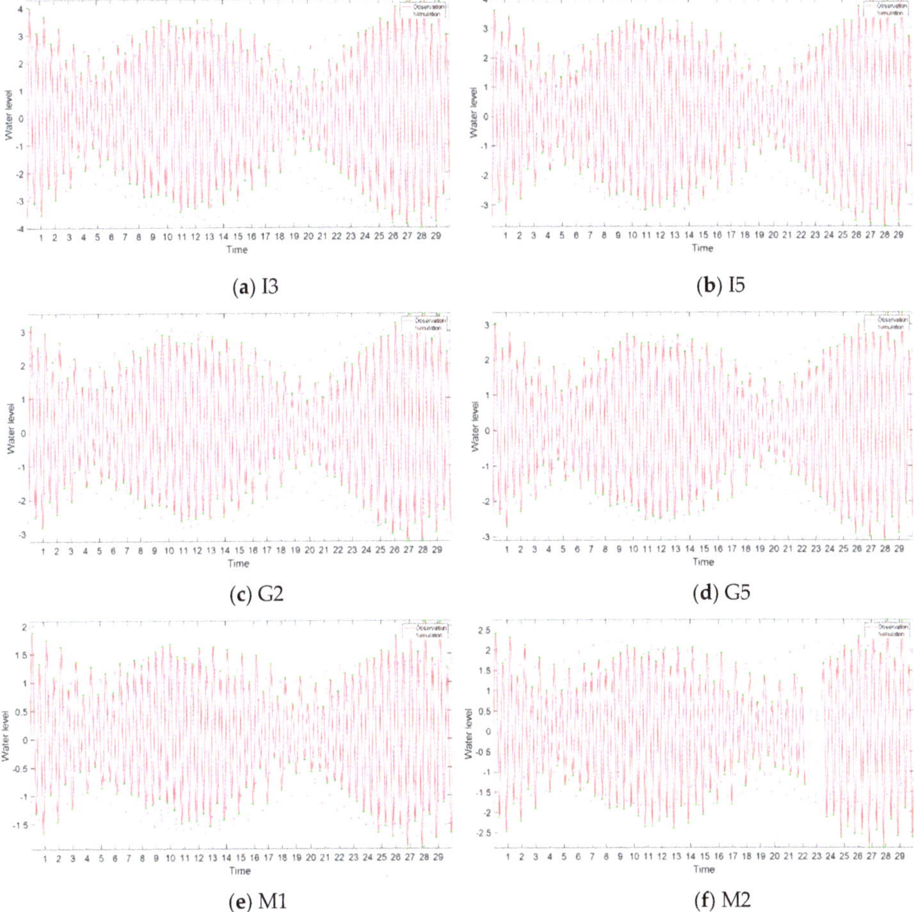

Figure 14. A comparison of water level between numerical results and observations at typical gauge stations; (**a**,**b**): water level at gauge station I3 and I5 (located in Incheon Region); (**c**,**d**): water level at gauge station G3 and G5 (located in Gunsan Region); (**e**,**f**): water level at gauge station M1 and M2 (located in Mokpo Region).

4. Conclusions

The main goal of this study was to use various numerical investigations to figure out the response of coastal tides in the WCK to the open boundary conditions and bottom roughness. After the application of three different open boundary conditions interpolated from three different assimilated tidal models (FES2014, NAO.99Jb and TPXO9.1), it has been shown that there were no significant differences between the responses of tidal amplitudes in the WCK induced by three open boundary conditions obtained from three assimilated tidal models. In addition, the numerical simulation of the tidal flow in the WCK should not use a uniform bottom roughness coefficient. Due to the complicated bathymetry, indented coastlines and bed variability of the WCK, it caused strong local effects on the tides in this region. Therefore, a non-uniform bottom roughness should be applied to the modeling whereby the smallest value can be applied for Incheon, a larger value for Mokpo, and the largest value for Gunsan. The largest value of the bottom roughness coefficient was applied to the Gunsan region because its bed form was more variable than other regions. The numerical results show that the accuracy of the modeling of the tidal elevation around the WCK was strongly dependent on the bottom roughness rather than the offshore tidal boundary conditions. Moreover, the numerical results can provide not only a better fit to the observations but also higher spatial resolutions in comparison to the results obtained from assimilated tidal models around the WCK.

However, it should be noted that the numerical results obtained from this study were still limited due to the coarse resolution (30 arcs/second) of bathymetry obtained from GEBCO2014, which was not sufficient to capture the real geometries, whose sizes were less than such resolution. Therefore, a further study with a higher resolution is necessary in order to obtain a more precise prediction of the tidal current and its elevation around the WCK. Furthermore, the wind forcing on the sea surface and the tidal energy dissipation should be taken into account.

Author Contributions: Conceptualization, V.T.N.; Data curation, V.T.N. and M.L.; Formal analysis, V.T.N.; Funding acquisition, V.T.N.; Investigation, V.T.N. and M.L.; Methodology, V.T.N.; Software, V.T.N. and M.L.; Supervision, V.T.N. All authors have read and agreed to the published version of the manuscript.

Funding: This work was funded by the National Research Foundation Grant of Korea (NRF-2018R1D1A1A09083747).

Acknowledgments: The authors also would like to thank the anonymous reviewers for their valuable and constructive comments to improve our manuscript.

Conflicts of Interest: The authors declare no conflict of interest.

Appendix A

Table A1. Validation of amplitude and phase obtained from three tidal models against observation data.

Tidal Constituent		RMSE			Range of Amplitudes and Phases
		FES2014 (1/16°)	NAO99Jb (1/12°)	TPXO9.1 (1/6°)	
M2	Amp (cm)	20.2	16.1	25	107.3 ÷ 295.8
	Phase (°)	8.2	5.4	6.6	36.8 ÷ 275.6
K1	Amp (cm)	4.2	4.9	6.2	21.7 ÷ 35.4
	Phase (°)	5.4	7.6	7.5	75.6 ÷ 193.1
S2	Amp (cm)	14.6	8.4	11.4	37.8 ÷ 114.1
	Phase (°)	10.9	8.8	7.1	59.6 ÷ 328.9
O1	Amp (cm)	3.8	4.8	3.9	16.3 ÷ 25
	Phase (°)	11	31.8	6.1	39 ÷ 142.9

Table A2. RMSE results obtained from uniform roughness bottom for FES2014.

Tidal Model	Constituent	Cd	RMSE (cm)			
			Mokpo (M)	Gunsan (G)	Incheon (I)	Mean
FES2014	M2	0.0015	10.17	34.83	40.18	36.30
		0.002	5.67	22.87	16.62	18.63
		0.0023	3.22	16.49	14.39	14.59
		0.0025	2.79	12.99	19.28	16.12
		0.003	6.24	10.58	35.57	26.63
	K1	0.0015	0.79	1.46	2.07	1.77
		0.002	0.98	1.30	2.03	1.70
		0.0023	1.16	1.48	2.34	1.96
		0.0025	1.29	1.70	2.64	2.21
		0.003	1.67	2.35	3.51	2.97
	S2	0.0015	2.63	10.94	12.21	11.15
		0.002	1.29	6.92	19.21	14.55
		0.0023	1.58	6.24	23.41	17.38
		0.0025	2.16	6.59	26.12	19.35
		0.003	3.76	8.99	31.90	23.77
	O1	0.0015	1.42	1.71	1.33	1.49
		0.002	1.82	2.38	1.86	2.07
		0.0023	1.99	2.67	2.26	2.40
		0.0025	2.10	2.86	2.53	2.63
		0.003	2.43	3.38	3.27	3.24

Table A3. RMSE results obtained from uniform roughness bottom for NAO99Jb.

Tidal Model	Constituent	Cd	RMSE (cm)			
			Mokpo (M)	Gunsan (G)	Incheon (I)	Mean
NAO99Jb	M2	0.0015	12.90	40.56	45.18	41.37
		0.002	7.38	26.96	19.39	21.89
		0.0023	4.14	19.93	14.16	16.06
		0.0025	2.29	15.75	17.64	16.07
		0.003	4.40	10.46	33.32	25.00
	K1	0.0015	0.95	1.65	2.17	1.90
		0.002	1.04	1.35	2.13	1.78
		0.0023	1.17	1.47	2.42	2.01
		0.0025	1.28	1.64	2.70	2.24
		0.003	1.64	2.30	3.60	3.01
	S2	0.0015	2.62	11.03	12.85	11.56
		0.002	1.53	7.07	19.89	15.05
		0.0023	1.82	6.39	24.02	17.84
		0.0025	2.37	6.77	26.70	19.78
		0.003	4.03	9.32	32.58	24.31
	O1	0.0015	0.49	0.84	1.49	1.21
		0.002	0.83	1.21	1.22	1.19
		0.0023	1.05	1.53	1.46	1.46
		0.0025	1.19	1.75	1.69	1.67
		0.003	1.54	2.28	2.37	2.27

Table A4. RMSE results obtained from uniform roughness bottom for TPXO9.1.

Tidal Model	Constituent	Cd	RMSE (cm)			
			Mokpo (M)	Gunsan (G)	Incheon (I)	Mean
TPXO9.1	M2	0.0015	10.49	34.68	40.07	36.19
		0.002	5.33	21.39	15.72	17.51
		0.0023	3.29	15.14	14.63	14.16
		0.0025	3.25	12.13	19.90	16.26
		0.003	7.42	11.13	36.85	27.64
	K1	0.0015	1.12	2.02	2.46	2.20
		0.002	0.99	1.36	1.91	1.65
		0.0023	1.01	1.21	1.93	1.62
		0.0025	1.06	1.25	2.10	1.74
		0.003	1.31	1.75	2.87	2.37
	S2	0.0015	2.17	10.22	12.63	11.13
		0.002	1.05	6.53	19.79	14.88
		0.0023	1.70	6.26	24.07	17.85
		0.0025	2.26	6.76	26.51	19.65
		0.003	4.07	9.54	32.34	24.17
	O1	0.0015	1.07	1.31	1.24	1.25
		0.002	1.46	1.95	1.55	1.71
		0.0023	1.70	2.33	1.98	2.10
		0.0025	1.80	2.50	2.25	2.31
		0.003	2.18	3.09	3.00	2.97

Table A5. List of simulation cases applying non-uniform bottom roughness coefficients to different sub-regions of WCK.

Case	Tide Model	Bottom Friction Coefficient			
		Mokpo	Gunsan	Incheon	Other Sub-Region
FES 2-1	FES2014	0.0025	0.0027	0.002	0.002
FES 2-2	FES2014	0.0025	0.003	0.002	0.002
FES 2-3	FES2014	0.0025	0.0035	0.0018	0.002
FES 2-4	FES2014	0.0025	0.0035	0.002	0.002
NAO 2-1	NAO99Jb	0.0025	0.0027	0.002	0.002
NAO 2-2	NAO99Jb	0.0025	0.003	0.002	0.002
NAO 2-3	NAO99Jb	0.0025	0.0035	0.0018	0.002
NAO 2-4	NAO99Jb	0.0025	0.0035	0.002	0.002
TPX 2-1	TPXO9.1	0.0025	0.0027	0.002	0.002
TPX 2-2	TPXO9.1	0.0025	0.003	0.002	0.002
TPX 2-3	TPXO9.1	0.0025	0.0035	0.0018	0.002
TPX 2-4	TPXO9.1	0.0025	0.0035	0.002	0.002

Table A6. RMSE results obtained from non-uniform roughness bottom for M2.

Tidal Constituent	Simulation Case	RMSE (cm)			
		Mokpo	Gunsan	Incheon	Mean
M2	FES 2-1	3.33	12.04	11.94	11.44
	FES 2-2	3.26	11.77	11.90	11.31
	FES 2-3	3.29	11.61	12.12	11.37
	FES 2-4	3.65	11.40	12.06	11.27
	NAO 2-1	2.25	15.00	12.75	13.09
	NAO 2-2	2.25	14.15	12.38	12.53
	NAO 2-3	2.27	13.39	13.83	13.00
	NAO 2-4	2.26	13.53	12.03	12.09
	TPX 2-1	4.10	11.09	11.91	11.08
	TPX 2-2	3.74	11.17	11.91	11.10
	TPX 2-3	3.88	11.18	11.68	10.97
	TPX 2-4	3.79	11.24	12.09	11.23

Table A7. RMSE results obtained from non-uniform roughness bottom for K1.

Tidal Constituent	Simulation Case	RMSE (cm)			
		Mokpo	Gunsan	Incheon	Mean
K1	FES 2-1	1.28	1.75	2.32	2.04
	FES 2-2	1.30	1.87	2.37	2.11
	FES 2-3	1.28	2.01	2.28	2.10
	FES 2-4	1.23	1.92	2.30	2.08
	NAO 2-1	1.28	1.72	2.42	2.09
	NAO 2-2	1.28	1.81	2.43	2.12
	NAO 2-3	1.28	1.96	2.36	2.13
	NAO 2-4	1.28	1.95	2.46	2.18
	TPX 2-1	1.04	1.32	1.90	1.63
	TPX 2-2	1.05	1.38	1.90	1.65
	TPX 2-3	1.05	1.50	1.84	1.66
	TPX 2-4	1.05	1.49	1.91	1.69

Table A8. RMSE results obtained from non-uniform roughness bottom for S2.

Tidal Constituent	Simulation Case	RMSE (cm)			
		Mokpo	Gunsan	Incheon	Mean
S2	FES 2-1	2.25	6.92	22.09	16.56
	FES 2-2	2.13	7.25	22.04	16.58
	FES 2-3	2.14	8.08	20.63	15.75
	FES 2-4	1.94	7.81	21.98	16.63
	NAO 2-1	2.38	7.03	22.60	16.94
	NAO 2-2	2.39	7.52	22.91	17.23
	NAO 2-3	2.37	8.29	21.31	16.27
	NAO 2-4	2.37	8.31	23.28	17.62
	TPX 2-1	2.38	7.10	22.36	16.78
	TPX 2-2	2.34	7.61	22.72	17.12
	TPX 2-3	2.30	8.40	21.09	16.14
	TPX 2-4	2.31	8.43	23.03	17.48

Table A9. RMSE results obtained from non-uniform roughness bottom for O1.

Tidal Constituent	Simulation Case	RMSE (cm)			
		Mokpo	Gunsan	Incheon	Mean
O1	FES 2-1	2.13	2.95	2.44	2.62
	FES 2-2	2.18	3.06	2.51	2.71
	FES 2-3	2.14	3.09	2.48	2.70
	FES 2-4	2.12	3.07	2.43	2.67
	NAO 2-1	1.21	1.82	1.62	1.67
	NAO 2-2	1.20	1.86	1.62	1.68
	NAO 2-3	1.22	1.97	1.62	1.73
	NAO 2-4	1.21	1.95	1.65	1.74
	TPX 2-1	1.91	2.70	2.22	2.39
	TPX 2-2	1.83	2.64	2.17	2.33
	TPX 2-3	1.82	2.71	2.16	2.36
	TPX 2-4	1.87	2.77	2.24	2.42

Table A10. Averaged interpolated amplitudes of tidal constituents at open boundaries.

Tidal Constituent	Ocean Tide Model	Mean Amplitude (cm)		
		A	B	C
M2	FES2014	188.7	55.5	21.3
	NAO99Jb	186.3	56.9	20.9
	TPXO9.1	191.1	55.8	20.5
K1	FES2014	27.4	21.5	3.8
	NAO99Jb	30.1	21.6	4.2
	TPXO9.1	26.5	21.4	3.4
S2	FES2014	55.4	23.8	10.6
	NAO99Jb	57.4	24.0	10.4
	TPXO9.1	52.0	23.9	10.5
O1	FES2014	22.0	16.8	4.2
	NAO99Jb	21.6	17.2	4.2
	TPXO9.1	22.9	17.0	3.3

Table A11. Averaged difference of interpolated amplitude on open boundaries (cm).

Tidal Constituent	Ocean Tide Model	RMSE (cm)			
		A	B	C	Mean
M2	FES2014	2.65	1.85	0.83	1.84
	TPXO9.1	5.20	2.78	1.09	2.91
K1	FES2014	2.72	0.41	0.78	0.91
	TPXO9.1	3.66	0.42	1.70	1.27
S2	FES2014	2.07	0.57	0.20	0.80
	TPXO9.1	5.43	0.91	0.41	1.77
O1	FES2014	0.50	0.53	0.55	0.53
	TPXO9.1	1.38	0.58	1.17	0.78

Table A12. D_a per 1cm increase of offshore tidal amplitude on boundary B.

Tidal Constituent	a_1-a_2	D_a			
		Incheon	Gunsan	Mokpo	Mean
M2	20–40	3.40	3.76	2.39	3.25
	40–60	2.51	3.04	1.80	2.62
	60–80	1.75	2.30	1.36	2.01
	80–100	1.60	2.15	1.26	1.91
K1	20–40	1.27	0.99	0.81	0.91
	40–60	1.51	1.20	0.97	1.09
	60–80	1.48	1.20	0.99	1.10
	80–100	1.30	1.07	0.91	1.00

Table A13. D_a per 1 cm increase of the tidal amplitude of K1 on the open boundary B.

Tidal Constituent		D_a			
		Incheon	Gunsan	Mokpo	Mean
M2	10–20	2.02	2.54	1.48	2.16
	20–30	1.80	2.33	1.36	1.96
K1	10–20	1.37	1.29	1.03	1.31
	20–30	1.37	1.29	1.04	1.31

Table A14. A comparison of water surface level at the gauge stations.

Sta. No.	Station Name	Water Surface Amplitude (m)					
		Neap Tide		Percentage Difference	Spring Tide		Percentage Difference
		Obs.	Sim.		Obs.	Sim.	
M1	Heuksando	1.18	0.9	24%	3.48	3.09	11%
M2	Mokpo	1.6	1.75	9%	4.99	4.04	19%
G1	Yeonggwang	2.06	2.43	15%	6.34	6.16	3%
G2	Wido	1.95	2.29	15%	6.2	5.95	4%
G3	Gunsan	2.08	2.64	21%	6.81	6.66	2%
G4	Janghang	2.19	2.8	22%	7.01	6.89	2%
G5	Eocheongdo	1.78	1.95	9%	5.85	5.28	10%
G6	Seocheonmaryang	2.13	2.54	16%	6.8	6.48	5%
G7	Boryeong	2.23	2.76	19%	7.15	6.14	14%
G8	Anheung	X(*)	2.29		6.46	5.86	9%
I1	Taean	2.2	2.9	24%	7.2	6.52	9%
I2	Pyeongtaek	2.8	2.62	7%	9	7.39	18%
I3	Daesan	2.35	3.18	26%	7.68	6.92	10%
I4	Ansan	X(*)	3.1		8.43	7.1	16%
I5	Gureopdo	2.16	2.57	16%	7.15	6.04	16%
I6	Yeongheungdo	2.54	3.41	26%	8.58	7.19	16%
I7	IncheonSongdo	2.15	3.08	30%	9.35	7.15	24%
I8	Incheon	2.23	3.26	32%	9.59	7.33	24%
I9	Yeongjongbridge	2.35	2.89	19%	9.87	7.03	29%
I10	Gyeongin	2.27	2.98	24%	8.68	7.08	18%
I11	Ganghwa	2.47	3.63	32%	7	5.73	18%

X(*): observation data were missing at this station.

References

1. Bao, X.W.; Gao, G.P.; Yan, J. Three-dimensional simulation of tide and tidal current characteristics in the East China Sea. *Oceanol. Acta* **2000**, *24*, 135–149. [CrossRef]
2. Choi, B.H. *A Tidal Model of the Yellow Sea and the East China Sea*; Rep. 20–02; Korea Ocean Research and Development Institute: Ansan, Korea, 1980.
3. Guo, X.; Yanagi, T. Three-dimensional structure of tidal current in the East China Sea and Yellow Sea. *J. Oceanogr.* **1998**, *54*, 651–668. [CrossRef]
4. Teague, W.J.; Perkins, L.H.T.; Hallock, Z.R.; Jacobs, G.A. Current and tide observations in the southern Yellow Sea. *J. Geophys. Res.* **1998**, *103*, 27783–27793. [CrossRef]
5. Le Fevre, F.; Le Provost, C.; Lyard, F.H. How can we improve a global ocean tide model at a regional scale? A test on the Yellow Sea and the East China Sea. *J. Geophys. Res.* **2000**, *105*, 8707–8725. [CrossRef]
6. Kim, K.O. Tidal Simulation in the Yellow and East China Seas by Finite Element Numerical Models. Master's, Thesis, Sunkyunkwan University, Suwon, Korea, 2000. (In Korean).
7. Naimie, C.E.; Blain, C.A.; Lynch, D.R. Seasonal mean circulation in the Yellow Sea—A model-generated climatology. *Cont. Shelf Res.* **2001**, *21*, 667–695. [CrossRef]
8. Suh, S.W. Reproduction of shallow tides and tidal asymmetry by using finely resolved grid on the west coast of Korea. *J. Korean Soc. Coast. Ocean. Eng.* **2011**, *23*, 313–325. (In Korean) [CrossRef]
9. Tang, Y. Numerical modeling of the tide-induced residual current in the East China Sea. *Prog. Oceanogr.* **1988**, *21*, 417–429.
10. Kantha, L.H.; Bang, I.; Choi, J.-K.; Suk, M.-S. Shallow water tides in the seas around Korea. *J. Korean Soc. Oceanogr.* **1996**, *31*, 123–133.
11. Kang, S.K.; Lee, S.-R.; Lie, H.-J. Fine grid tidal modeling of the Yellow and East china seas. *Cont. Shelf Res.* **1998**, *18*, 739–772. [CrossRef]
12. Lee, H.J.; Jung, K.T.; So, J.K.; Chung, J.Y. A three-dimensional mixed finite-difference Galerkin function model for the oceanic circulation in the Yellow Sea and the East China Sea in the presence of M2 tide. *Cont. Shelf Res.* **2002**, *22*, 67–91. [CrossRef]
13. Ogura, S. The tides in the seas adjacent to Japan. *Bull. Hydrogr. Dep. Imp. Jpn. Navy* **1933**, *7*, 1–189.

14. An, H.S. A numerical experiment of M2 tide in the Yellow Sea. *J. Oceanogr. Soc. Jpn.* **1977**, *33*, 103–110. [CrossRef]
15. Nishida, H. Improved tidal charts for the western part of the North Pacific Ocean. *Rep. Hydrogr. Res.* **1980**, *15*, 55–70.
16. Yanagi, T.; Inoue, K. Tide and tidal current in the Yellow Sea/East China Sea. *Lamer* **1994**, *32*, 153–165.
17. Lee, J.C.; Jung, K.T. Application of eddy viscosity closure models for M2 tide and tidal currents in the Yellow Sea and the East China Sea. *Cont. Shelf Res.* **1999**, *19*, 445–475. [CrossRef]
18. Lee, M.E.; Kim, G.; Nguyen, V.T. Effect of local refinement of unstructured grid on the tidal modeling in the south-western coast of Korea. *J. Coast. Res.* **2013**, *65*, 2017–2022. [CrossRef]
19. Schwiderski, E.W. On Charting Global Ocean Tides. *Rev. Geophys.* **1980**, *18*, 243–268. [CrossRef]
20. Matsumoto, K.; Takanezawa, T.; Ooe, M. Ocean tide models developed by assimilating TOPEX/POSEIDON altimeter data into hydrodynamical model: A global model and a regional model around Japan. *J. Oceanogr.* **2000**, *56*, 567–581. [CrossRef]
21. Kantha, L.H. Barotropic tides in the global oceans from a nonlinear tidal model assimilating altimetric tides 1. Model description and results. *J. Geophys. Res.* **1995**, *100*, 25283–25308. [CrossRef]
22. Le Provost, C.; Genco, M.L.; Lyard, F.; Vincent, P.; Canceil, P. Spectroscopy of the world ocean tides from a finite element hydrodynamic model. *J. Geophys. Res.* **1994**, *99*, 24777–24797. [CrossRef]
23. Han, G.; Hendry, R.; Ikeda, M. Assimilating TOPEX/POSEIDON derived tides in a primitive equation model over the Newfoundland Shelf. *Cont. Shelf Res.* **2000**, *20*, 83–108. [CrossRef]
24. Lyard, F.; Lefevre, F.; Letellier, T.; Francis, O. Modelling the global ocean tides: A modern insight from FES2004. *Ocean. Dyn.* **2006**, *56*, 394–415. [CrossRef]
25. Fu, L.L.; Christensen, E.J.; Yamarone, C.A., Jr.; Lefebvre, M.; Menard, Y.; Dorrer, M.; Escudier, P. TOPEX/POSEIDON mission overview. *J. Geophys. Res.* **1994**, *99*, 24369–24381. [CrossRef]
26. Dronkers, J.J. *Tidal Computation in Rivers and Coastal Waters*; North-Holland Publishing Company: Amsterdam, The Netherlands, 1964; pp. 3–87.
27. Brière, C.; Abadie, S.; Bretel, P.; Lang, P. Assessment of TELEMAC system performances, a hydrodynamic case study of Anglet, France. *Coast. Eng.* **2007**, *54*, 345–356. [CrossRef]
28. Villaret, C.; Hervouet, J.M.; Kopmann, R.; Merkel, U.; Davies, A.G. Morphodynamic modelling using the Telemac finite element system. *Comput. Geosci.* **2013**, *53*, 105–113. [CrossRef]
29. Choi, B.C.; Ko, J.S. Modelling of tides in the East Asian Marginal Seas. *J. Korean Soc. Ocean. Eng.* **1994**, *6*, 94–108.
30. Blain, C.A. Development of a data sampling strategy for semi-enclosed seas using a shallow-water model. *J. Atmos. Ocean. Technol.* **1997**, *14*, 1157–1173. [CrossRef]
31. Zhao, B.; Fang, G.; Chao, D. Numerical simulations of the tide and tidal current in the Bohai Sea, the Yellow Sea and the East China Sea. *Acta Oceanol. Sin.* **1994**, *16*, 1–10.
32. Ye, A.; Mei, L. Numerical modeling of tidal waves in the Bohai Sea, the Huanghai Sea and the East China Sea. *Oceanol. Limnol. Sin.* **1995**, *26*, 63–70. (In Chinese with English abstract)
33. Defant, A. *Physical Oceanography*; Pergamon Press: Oxford, UK, 1960; Volume II.

Article

Numerical and Experimental Study of Classical Hydraulic Jump

Eugene Retsinis * and Panayiotis Papanicolaou

School of Civil Engineering, National Technical University of Athens, 9 Iroon Polytexneiou St., 15780 Athens, Greece; panospap@mail.ntua.gr

* Correspondence: retsinis76@hotmail.com; Tel.: +30-6944-916-187

Received: 20 May 2020; Accepted: 19 June 2020; Published: 21 June 2020

Abstract: The present work is an effort to simulate numerically a classical hydraulic jump in a horizontal open channel with a rectangular cross-section, as far as the jump location and free surface elevation is concerned, and compare the results to experiments with Froude numbers in the range 2.44 to 5.38. The governing equations describing the unsteady one-dimensional rapidly varied flow have been solved with the assumption of non-hydrostatic pressure distribution. Two finite difference schemes were used for the discretization of the mass and momentum conservation equations, along with the appropriate initial and boundary conditions. The method of specified intervals has been employed for the calculation of the velocity at the downstream boundary node. Artificial viscosity was required for damping the oscillations near the steep gradients of the jump. An iterative algorithm was used to minimize the difference of flow depth between two successive iterations that must be less than a threshold value, for achieving steady state solution. The time interval varied in each iteration as a function of the Courant number for stability reasons. Comparison of the numerical results with experiments showed the validity of the computations. The numerical codes have been implemented in house using a Matlab® environment.

Keywords: hydraulic jump; Boussinesq equations; MacCormack; Dissipative two-four; method of specified intervals

---

1. Introduction

The flow in prismatic open channels as a civil engineering specialty has been studied for over 100 years via laboratory experiments, theoretical one-dimensional analysis, as well as with numerical modeling. The flow may be characterized as a gradually varied (GVF) or a rapidly varied (RVF) flow if changes of flow depth are gradual over a significant length or abrupt over a short length of the channel respectively. A characteristic dimensionless parameter describing the type of flow is the Froude number, $\mathrm{Fr} = \mathrm{u}/\sqrt{\mathrm{gh}}$ that expresses the ratio of inertia over the gravity forces, u being the average, over a cross section velocity; g is the gravitational acceleration and D is the hydraulic depth, i.e., the depth of the equivalent orthogonal channel with bottom width equal to the width of free surface of the prismatic channel. When inertia forces are dominant, $\mathrm{Fr} > 1$ and the flow is called supercritical, otherwise when $\mathrm{Fr} < 1$, the flow is named subcritical. Critical flow ($\mathrm{Fr} = 1$) may occur, but it is quite unstable, oscillating between a subcritical and a supercritical state. A special case of a RVF is the hydraulic jump, which is formed in an open channel when supercritical flow turns into subcritical, resulting in an abrupt rise of the free surface, formation of a surface roller and high energy dissipation locally. The hydraulic jump in civil engineering applications can be used as an energy dissipation mechanism, for mixing purposes as well as for the aeration of the flow.

Hydraulic jumps appear usually in stilling basins for energy dissipation of spillways when required. The sequent (subcritical) depth, the length (distance between subcritical and supercritical flow) and the location where the jump occurs are important parameters for the safe design of such structures. Among the experiments performed to determine these parameters are the pioneering works [1,2], where the one-dimensional momentum equation was derived relating the sequent depth ratio to the upstream Froude number. Moreover, in more recent experiments on hydraulic jumps [3–7], basic flow variables have been determined such as the sequent depth ratio, the length of the jump and the geometry of the surface roller.

The one-dimensional depth-averaged equations of mass conservation (continuity) and momentum (Navier–Stokes), describing the unsteady, rapidly varied open channel flow with non-hydrostatic pressure distribution, form a two-equation system named the Boussinesq equations. Assuming parallel flow and hydrostatics over the depth pressure distribution, the simplified system is called the Saint Venant (shallow water) equation. In both systems of equations the variables are the depth of flow and the average over depth velocity. A hydraulic jump can be modeled using the Boussinesq equations, which is the main objective of this work. So far, the shallow water Saint Venant equations have been used in several investigations for the numerical simulation of the classical hydraulic jump. For example, the finite element method was used [8] to solve the Saint Venant equations up to the point that steady state was reached. The assumption of hydrostatic pressure distribution employed is not valid in the region of the hydraulic jump, because the substantial curvature of the streamlines results in vertical accelerations that cannot be neglected. Moreover, the energy losses were neglected (weak hydraulic jump), making questionable the validity of simulation results for practical purposes. An attempt was made [9], to combine the knowledge regarding vorticity generation in a hydraulic jump, using the Navier–Stokes and shallow water equations. The Saint Venant equations were also modified [10], including terms related to turbulent shear stress and non-uniform velocity distribution, applying a depth-averaged approach. The equations were solved by the finite element method for the location and the profile of a hydraulic jump with upstream Froude numbers in the range 2 to 7. An effort to simulate the classical hydraulic jump using the shallow water equations and solve them using the MacCormack numerical scheme [11], produced results that compare well with the theoretical ones. The finite volume method [12] has been used to solve the governing shallow water equations. This model was applied to a hydraulic jump, and the numerical results are in agreement with experiments.

The numerical solution of Navier–Stokes equations has been widely used in modeling of the hydraulic jump. The use of the Reynolds-Averaged Navier–Stokes (RANS) approach and the k-ε turbulence closure model [13], was applied to study the turbulence characteristics of the hydraulic jump, which had to be adapted to flow cases of a moving free surface by means of the partially explicit approach. The study includes both free and forced hydraulic jumps with Froude numbers in the range from 2.1 to 7.6. Comparison of the numerical predictions with velocity measurements using one-dimensional Laser Doppler Anemometry, showed that the results qualitatively compare well, regarding the mean velocity and turbulence intensity in the flow direction. RANS equations were solved using a turbulence closure k-ε model [14], to study the hydraulic jump in a straight horizontal channel with Froude numbers 2.0 and 4.0. The mixed Eulerian-Lagrangian method for the calculation of the free surface of the flow is utilized to overcome the problem posed by a moving free boundary. A detailed study of the internal and external characteristics of a hydraulic jump was made and compared to experiments where possible. The surface roller and recirculation zone are found to play a dominant role in turbulence generation and dissipation. The study of free and submerged hydraulic jumps in a rectangular open channel downstream a sluice gate [15] has also been considered. Air-water two-phase flows are considered in the numerical simulations using RANS and turbulence closure k-ε modeling [15], and compared to velocity measurements made using Acoustic Doppler and Particle Image Velocimetry. The numerical results concerning water depths, the hydraulic jump length and the velocity profiles compare well with laboratory measurements. A three-dimensional model was implemented [16], using OpenFOAM software to analyze a hydraulic jump with upstream Froude number 6.1 in a horizontal, smooth, rectangular open channel. RANS equations were solved using three turbulence closure models, namely the Standard k-ε, the Re-Normalization Group (RNG) k-ε, and the Shear Stress Transport (SST) k-ω model. Direct Numerical Simulation (DNS) was used [17], in a hydraulic jump with Froude number equal 2.0, using the volume of fluid method for tracking the free surface. Results have been produced regarding the mean velocity field, Reynolds stresses, turbulence production and dissipation, velocity spectra and air entrainment concentration. Finally the two recent works [18,19] present a very thorough investigation on datasets for numerical modeling assessment of the hydraulic jump.

In summary, the shallow water one-dimensional equations are capable of modeling only some characteristics of a transient hydraulic jump, while RANS combined with turbulence closure models or DNS can capture the turbulent structure of a steady hydraulic jump more accurately. Implementation of such algorithms though is limited by the Reynolds number of the flow that usually exceeds 10^6. Moreover, the computational cost of such models is very high, because the computational time if a decent computing machine is very long. For practical, mainly civil engineering applications, shallow water modeling is much simpler to use, while the results can be acceptable. In the present work, we will show that one-dimensional modeling of a hydraulic jump is improved using the more accurate, unsteady, one-dimensional Boussinesq equations. They are solved numerically using the MacCormack and the Dissipative two-four finite difference schemes. The practicality of such a model used in applications like the design of stilling basins, has led us to choose this model for analysis. The maximum error of the flow depth is used as an iterative parameter, for achieving the steady state solution for the free surface profile and the location of a hydraulic jump, in a horizontal rectangular open channel for a wide range of Froude numbers. To apply the Boussinesq model at the downstream boundary node, the method of specified intervals was used for the evaluation of the unknown flow variables. The results are compared to experiments performed at the Laboratory of Applied Hydraulics, School of Civil Engineering, National Technical University of Athens, Greece, showing that such numerical schemes can accurately simulate the classical hydraulic jump.

2. Governing Equations

The equations modeling unsteady one-dimensional rapidly varied flows are the Boussinesq equations [20]. These include additional terms for the non-hydrostatic pressure distribution due to curved streamlines. The basic assumptions made are: (1) the vertical velocity varies from zero at the channel bottom to its maximum value at the free surface, (2) the velocity in the main flow direction is uniformly distributed over the depth, (3) the velocity in the lateral direction is zero, (4) the fluid is incompressible, (5) the channel is prismatic of a rectangular cross section with rigid bottom and sides, (6) the longitudinal bottom slope is small, and (7) the formulas regarding energy friction slope for steady flow can be used for the unsteady flow as well. The one-dimensional Boussinesq equations for mass and momentum conservation are written in vector form as:

$$\frac{\partial G}{\partial t} + \frac{\partial F}{\partial x} = S, \tag{1}$$

where:

$$G = \begin{bmatrix} h \\ uh \end{bmatrix}, F = \begin{bmatrix} uh \\ u^2h + \left(\frac{1}{2}\right)gh^2 - \left(\frac{1}{3}\right)h^3E \end{bmatrix}, S = \begin{bmatrix} 0 \\ gh(S_o - S_f) \end{bmatrix} \tag{2}$$

$$E = \frac{\partial^2 u}{\partial x \partial t} + u\frac{\partial^2 u}{\partial x^2} - \left(\frac{\partial u}{\partial x}\right)^2, \tag{3}$$

x is the longitudinal distance along the channel bottom as shown in Figure 1, t is the time, h = h(x,t) and u = u(x,t) are the unknown flow depth and mean velocity components in the main flow direction respectively, S_f is the energy grade line slope, S_o is the longitudinal bottom slope and g is the gravitational acceleration. E = E(x,t) is the Boussinesq term, which makes the difference if compared to the Saint Venant equations where the pressure distribution is assumed to be hydrostatic. The energy grade slope can be computed using the Manning formula in SI, $u = (R^{2/3}S^{1/2})/n$, where u is the mean over the wetted cross-section velocity, R is the hydraulic radius (cross sectional area over the wetted perimeter ratio), $S = S_f$ the energy grade slope and n the Manning friction coefficient as $S_f = n_f^2 u^2/R^{4/3}$. Alternatively, the Darcy–Weisbach formula could be used for computing S_f, but we used the Manning formula in order to simplify calculations since the friction coefficient is kept constant for every step and the results are not affected.

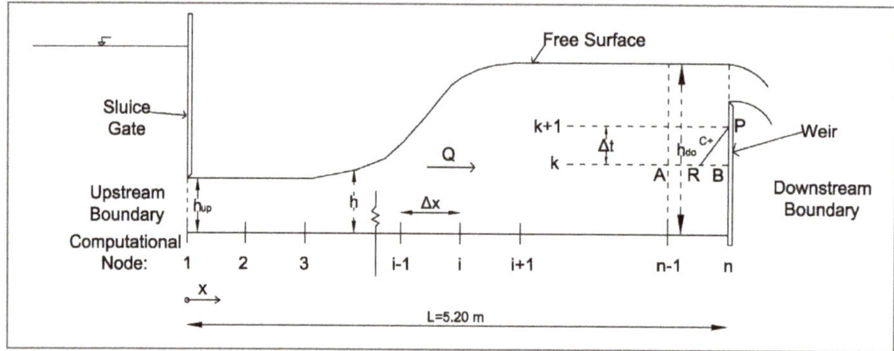

Figure 1. Physical and computational domain of the classical hydraulic jump.

3. Numerical Schemes

The classical hydraulic jump Boussinesq equations cannot be solved analytically, therefore a numerical method has to be implemented. In this work, the MacCormack [21] and the Dissipative two-four [22] finite difference schemes have been employed for the solution of the flow equations with the appropriate initial and boundary conditions. The first scheme is second order accurate both in space and time, and the second one is fourth order accurate in space and second order in time, allowing for proper simulation of the Boussinesq terms. The developed algorithm iterates until the change of the flow depth between two successive iterations is less than a fixed convergence value. Then the jump forms a part of the steady state solution.

A sketch of the jump with the computational grid is shown in Figure 1. The hydraulic jump is formed combining the use of a sluice gate upstream with a weir downstream. The origin of the spatial coordinate (x = 0) is set at the sluice gate location, while x is the longitudinal distance along the channel bottom measured from the origin. The distance between the sluice gate and the weir is L (equal to 5.20 m, the region of interest where the computational solution is sought), and it is discretized by a total number of n nodes including the boundary nodes, thus creating a uniform grid of points at distance $\Delta x = L/(n-1)$ in between. The index i denotes a spatial node, while the upstream and downstream boundaries correspond to nodes i = 1 and i = n with flow depths denoted as h_{up} and h_{do} respectively. A brief presentation of the numerical schemes for the discretization of Equation (1) follows.

3.1. MacCormack Scheme

This scheme is a two-step algorithm. For the spatial derivatives of Equation (1) forward finite differences are used in the predictor step and backward finite differences in the corrector step including in the computational stencil two spatial nodes in the predictor step, the nodes i + 1, i and in the corrector step the nodes i and i − 1 as follows:

Predictor step:

$$G_i^* = G_i^k - \lambda\left(F_{i+1}^k - F_i^k\right) + \Delta t S_i^k, \tag{4}$$

Corrector step:

$$G_i^{**} = G_i^* - \lambda\left(F_i^* - F_{i-1}^*\right) + \Delta t S_i^*, \tag{5}$$

The flow variables at the next iteration level k+1, and grid point i, are given by:

$$G_i^{k+1} = \frac{1}{2}\left(G_i^k + G_i^{**}\right), \tag{6}$$

where λ = Δt/Δx, Δt being the time interval and the superscripts k and k + 1 refer to the two successive levels of iteration. All variables with a single asterisk (*) refer to those computed at the predictor step where all variables with double asterisk (**) refer to those computed at the corrector step. The Boussinesq terms are neglected in this scheme.

3.2. Dissipative Two-Four Scheme

The Dissipative two-four scheme consists of a predictor and a corrector step, again for the spatial derivatives of Equation (1) forward finite differences are used in the predictor step and backward finite differences having in the corrector step, as including in the computational stencil, three spatial nodes in the predictor step, the nodes i + 2, i + 1 and i and in the corrector step the nodes i, i − 1 and i − 2:

Predictor step:

$$G_i^* = G_i^k + \frac{\lambda}{6}\left(F_{i+2}^k - 8F_{i+1}^k + 7F_i^k\right) + \Delta t S_i^k, \tag{7}$$

Corrector step:

$$G_i^{**} = \frac{1}{2}\left(G_i^k + G_i^*\right) + \frac{\lambda}{12}\left(-7F_i^* + 8F_{i-1}^* - F_{i-2}^*\right) + \frac{1}{2}\Delta t S_i^*, \tag{8}$$

The vector G_i^{k+1} at the next iteration level k + 1 and grid point i, is given by:

$$G_i^{k+1} = \frac{1}{2}\left(G_i^k + G_i^{**}\right), \tag{9}$$

Regarding Boussinesq terms, the second order derivative $\partial^2 u/\partial x^2$ is approximated using a three point central finite difference in both steps, while for the first order derivative $\partial u/\partial x$, forward finite difference is used in the predictor step, and a backward finite difference in the corrector step. The partial derivative $\partial^2 u/\partial x \partial t$ in the Boussinesq terms is ignored, since it is zero at steady state. Regarding the code implementation, Equations (4)–(9), have to be written in non-vector form.

Due to computational reasons in the case of the Dissipative two-four scheme, the flow equations are discretized using Equations (7) and (8) at the interior nodes, i.e., for i = 4, 5, ..., n−2, the Boussinesq terms are either considered or neglected, while at nodes i = 2, 3, n−1, the MacCormack scheme is used for discretizing Equations (4) and (5) and omitting the Boussinesq terms. Apparently, this is allowed since the jump appears far from the boundary nodes, therefore the Boussinesq terms are not significant since the flow is parallel.

3.3. Initial and Boundary Conditions

The appropriate initial and boundary conditions must be set in a well-posed problem. The number of characteristics that enter the computational domain determine the auxiliary conditions to be specified [23]. The initial condition includes steady, supercritical, gradually varied flow for the entire length of the channel, resulting in two characteristic curves entering the computational domain, thus two flow parameters must be specified at each grid point, the flow depth and the velocity. The depth at each grid point can be computed by numerical integration of the first order ordinary differential equation:

$$\frac{dh}{dx} = \frac{S_o - S_f}{1 - Fr^2}, \tag{10}$$

of the steady state gradually varied flow, Fr being the Froude number, $Fr = u/\sqrt{gh}$. The upstream depth h_{up} is known (obtained from the experiments downstream of the sluice gate) and the computation is performed with a Kutta-Merson method, beginning the calculations with the flow depth h_{up}.

The flow conditions are fixed at the two boundaries of the channel. For a given discharge and settings of the sluice gate upstream and the thin crested weir downstream, the flow depths are known and take constant values h_{up} and h_{do} at nodes i = 1 and i = n respectively. The mean velocity is known at i = 1 but it has to be computed at end node i = n. Apparently, the flow is assumed to be supercritical at node i = 1 and subcritical at node i = n.

The velocity at the end node i = n will be computed using the method of specified intervals and the positive characteristic equation discretized by finite differences. From Figure 1, points A and B correspond to nodes n−1 and n respectively at the known time level k, while the positive characteristic through point P with the unknown velocity at the downstream boundary at time level k+1 is indicated. The method of specified intervals is used to compute the velocity, the celerity and the flow depth at point R, which is the intersection of the positive characteristic through point P with the grid line of the known time level k using the following relationships [24]:

$$u_R = \frac{u_B + \lambda(c_B u_A - c_A u_B)}{1 + \lambda(u_B - u_A + c_B - c_A)}, \tag{11}$$

$$c_R = \frac{c_B + \lambda u_R (c_B - c_A)}{1 + \lambda(c_B - c_A)}, \tag{12}$$

$$h_R = h_B - \lambda(u_R + c_R)(h_B - h_A), \tag{13}$$

where $c = \sqrt{gh}$ is the celerity of a small amplitude wave propagating inside a rectangular open channel in shallow water. The energy grade line slope at point R is estimated from the following Equation:

$$S_{f_R} = n^2 u_R^2 / R_R^{4/3}, \tag{14}$$

Then the velocity u_n^{k+1} at point P can be computed from the following relationship:

$$u_P = u_n^{k+1} = u_R^k - \frac{g}{c_R^k}\left(h_n^{k+1} - h_R^k\right) + g\Delta t\left(S_o - S_{fR}^k\right), \tag{15}$$

3.4. Courant Condition Artificial Viscosity

The variable time step Δt in each iteration is restricted according to the Courant criterion for stability purposes, and it is computed from:

$$\Delta t = \frac{c_n \Delta x}{\max\left(|u| + \sqrt{gh}\right)}, \tag{16}$$

where c_n is the Courant number that must be less than or equal to 0.65 [24], and Δx is the constant spatial step as shown in Figure 1.

To dampen the oscillations that occur around the jump region, artificial viscosity has to be added to the numerical schemes. The formulas used are those proposed by [25].

First the parameter ξ_i at the computational node i is calculated:

$$\xi_i = \frac{\Delta x}{\Delta t} \frac{|h_{i+1} - 2h_i + h_{i-1}|}{|h_{i+1}| + 2|h_i| + |h_{i-1}|}, \quad \text{for the nodes, } i = 2, \ldots, n-1 \tag{17}$$

$$\xi_i = \frac{\Delta x}{\Delta t} \frac{|h_{i+1} - h_i|}{|h_{i+1}| + |h_i|}, \quad \text{for the node, } i = 1 \tag{18}$$

$$\xi_i = \frac{\Delta x}{\Delta t}\frac{|h_i - h_{i-1}|}{|h_i| + |h_{i-1}|}, \text{ for the node, } i = n \tag{19}$$

Then at the center of segment i + 1/2, between nodes i and i + 1:

$$\xi_{i+(1/2)} = k_{art}\max(\xi_i, \xi_{i+1}), \tag{20}$$

Similarity between nodes i − 1 and i:

$$\xi_{i-(1/2)} = k_{art}\max(\xi_{i-1}, \xi_i), \tag{21}$$

where k_{art} is a coefficient adjusting the amount of dissipation.

Finally the flow variables h and u, at iteration level k+1 are modified to the new ones according to the following relationships:

$$f_{new}{}_i^{k+1} = f_{old}{}_i^{k+1} + \xi_{i+(1/2)}\left(f_{old}{}_{i+1}^{k+1} - f_{old}{}_i^{k+1}\right) - \xi_{i-(1/2)}\left(f_{old}{}_i^{k+1} - f_{old}{}_{i-1}^{k+1}\right), \tag{22}$$

where the variable f stands either for the flow depth or for the velocity. The numerical codes have been implemented in house using Matlab® computational environment, while the input data for the numerical simulations are the channel geometry, the flow depths h_{up}, h_{do} and the flow rate Q.

4. Results

The experiments were performed at the Laboratory of Applied Hydraulics of the School of Civil Engineering at the National Technical University of Athens, Greece, in a 10.50 m long flume with a rectangular cross section 0.248 m wide and 0.50 m deep. The channel has a steel bottom and vertical side walls made of glass. The water supply system consists of a constant head tank, a pipeline and a regulating valve to adjust the flow rate that was measured with a Venturi meter and did not exceed 50 L/s. The flow was controlled by a sluice gate setting the supercritical flow conditions upstream, and a thin crested weir at the end of the channel located 5.20 m downstream from the sluice gate for controlling the tailwater, subcritical flow. The hydraulic jump formed at some location between the sluice gate and the weir, depending upon the upstream, supercritical and tailwater, subcritical depths. The flow depths and free surface profile of the jumps were measured by a point gauge with accuracy ±0.1 mm.

Six experiments were implemented, the flow conditions of which are shown in Table 1. In the table appear the flow rate q, the supercritical upstream and subcritical downstream depths h_{up} and h_{do} respectively, and the Froude number of the flow, Fr, at the toe of the jump from the experimental measurements. The six different jump cases measured, were computed using same conditions (upstream and tailwater depths and the flow rate) and the numerical results are compared to experiments in the following paragraphs.

Table 1. Experimental flow variables of the hydraulic jumps.

Test Case/Experiment	q (L/s/m)	Fr	h_{up} (m)	h_{do} (m)
1	24.28	2.44	0.0217	0.0682
2	54.32	3.06	0.0319	0.1335
3	24.28	3.40	0.0174	0.0735
4	28.72	4.03	0.0174	0.0788
5	28.72	4.48	0.0162	0.0906
6	21.72	5.38	0.0119	0.0895

For the numerical modeling of the experiments shown in Table 1 three discretizing schemes of the Boussinesq equations were used: (i) the Dissipative two-four scheme with the influence of Boussinesq terms, (ii) the Dissipative two-four scheme without the influence of non-hydrostatic pressure distribution terms, and (iii) the MacCormack scheme without Boussinesq terms. The time interval computed from Equation (16) varied in every iteration. Moreover, Equation (22) was applied to dampen the oscillations near the jump. In all test cases 100 spatial nodes were used for discretization (n = 100), resulting in $\Delta x = 0.0525$ m with acceptable spatial resolution. All the results regarding the computed dimensionless flow depth h/h_{do}, alongside the dimensionless longitudinal distance of the channel, x/L, for each test case are shown in Figures 2a, 3a, 4a, 5a, 6a and 7a. The measured and computed flow profiles by the three numerical algorithms are plotted together along the channel, and the significance of the Boussinesq terms (due to non-parallel streamlines) inside the region of the hydraulic jump are depicted sideways in Figures 2b, 3b, 4b, 5b, 6b and 7b. Outside the jump these terms are almost zero (small enough), thus confirming the hydrostatic pressure distribution. From the numerical results, it must be noted that the flow depth increases from the vena contracta downstream of the sluice gate up to the jump, due to energy losses encountered in the governing equations. One may note that in test case 1, the dissipative behavior of the numerical results is probably due to the artificial dissipation added in the numerical scheme. In addition, from the Boussinesq term distribution along the channel, one may note that this term is not zero at the upstream end, therefore we would not expect a smooth transition in free surface from supercritical to subcritical flow. Furthermore, note that the Froude number is quite low and the hydraulic jump is characterized as weak [26], since $1.7 < Fr < 2.5$, which means that it is a non-fully developed jump because of the weak surface roller with low energy loss. From these figures we conclude that the agreement between experiments and computations is satisfactory.

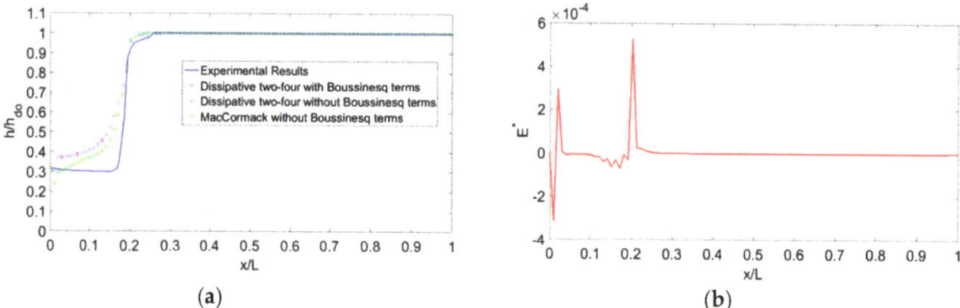

Figure 2. Experimental and Numerical Results for test case 1: (**a**) Free surface profile; (**b**) Boussinesq term versus distance.

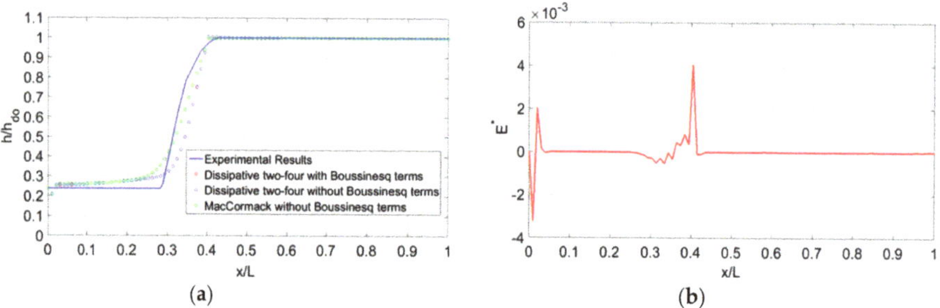

Figure 3. Experimental and Numerical Results for test case 2: (**a**) Free surface profile; (**b**) Boussinesq term versus distance.

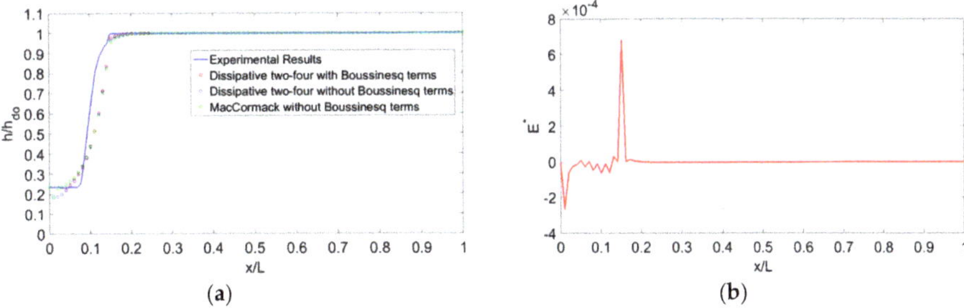

Figure 4. Experimental and Numerical Results for test case 3: (**a**) Free surface profile; (**b**) Boussinesq term versus distance.

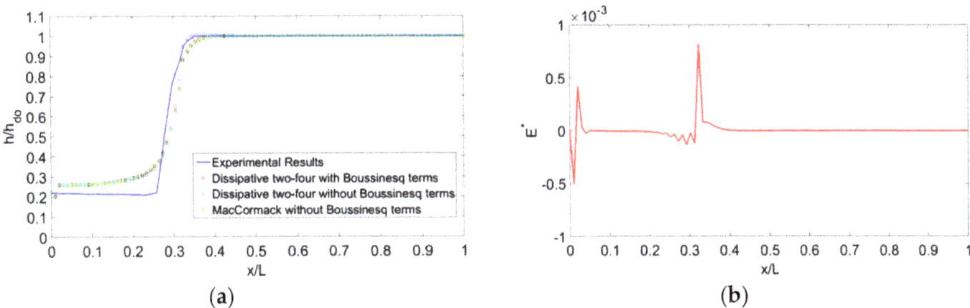

Figure 5. Experimental and Numerical Results for test case 4: (**a**) Free surface profile; (**b**) Boussinesq term versus distance.

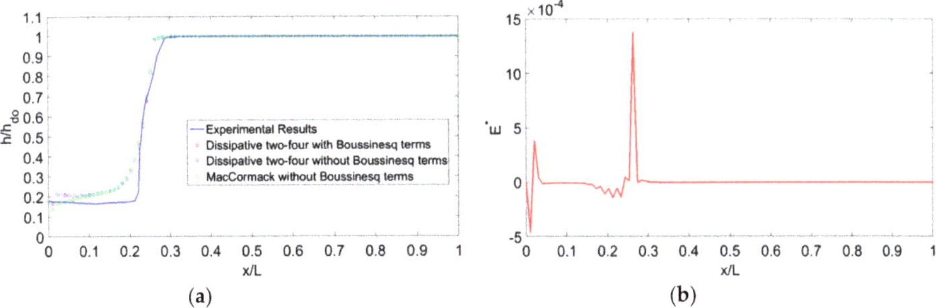

Figure 6. Experimental and Numerical Results for test case 5: (**a**) Free surface profile; (**b**) Boussinesq term versus distance.

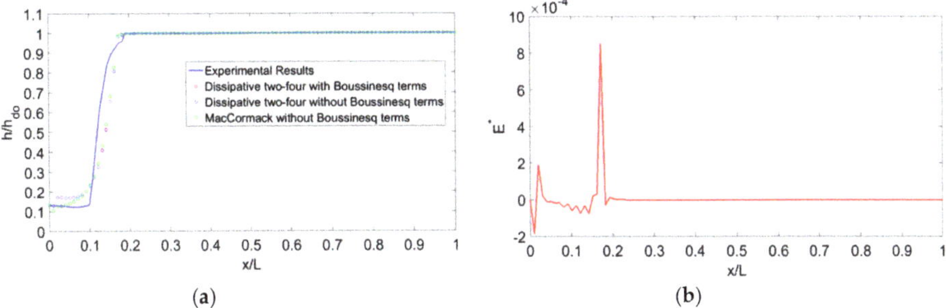

Figure 7. Experimental and Numerical Results for test case 6: (**a**) Free surface profile; (**b**) Boussinesq term versus distance.

Figure 8a,b exhibit the variation of the dimensionless spatially (over a cross-section) averaged flow velocity in the main flow direction x, u/u_{up}, where u_{up} is the average velocity at x = 0, with the dimensionless longitudinal distance, x/L, for all three algorithms for test cases 1 and 6 respectively. It is evident that the MacCormack scheme overestimates the flow velocity at the upstream end of the channel, while the results from the other two methods are almost identical. Figure 9a,b depict the evolution of convergence criteria of the maximum absolute change in flow depth between two successive iterations, until a steady state is obtained for test case 2 using the Dissipative two-four scheme with Boussinesq terms, and test case 5 using the MacCormack scheme without Boussinesq terms respectively. Figures 10 and 11 present the temporal evolution in "computational-pseudo time" of the free surface elevation for the Dissipative two-four scheme including the Boussinesq terms for test cases 3 and 4 respectively, both in dimensionless form. Note that the jump moves upstream until it is stabilized. Similar results have been produced for all other test cases.

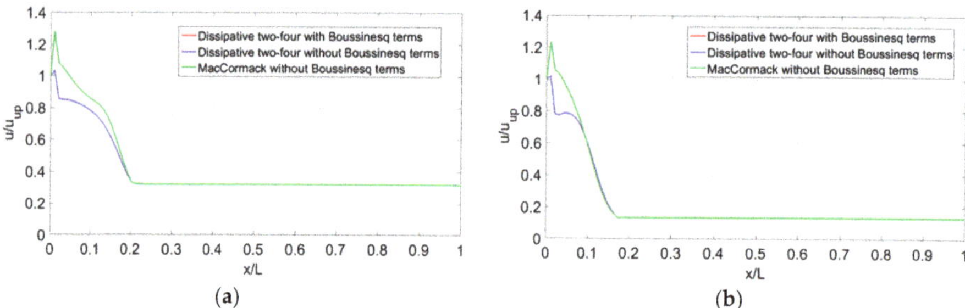

Figure 8. Variation of the mean velocity alongside the channel: (**a**) for test case 1; (**b**) for test case 6.

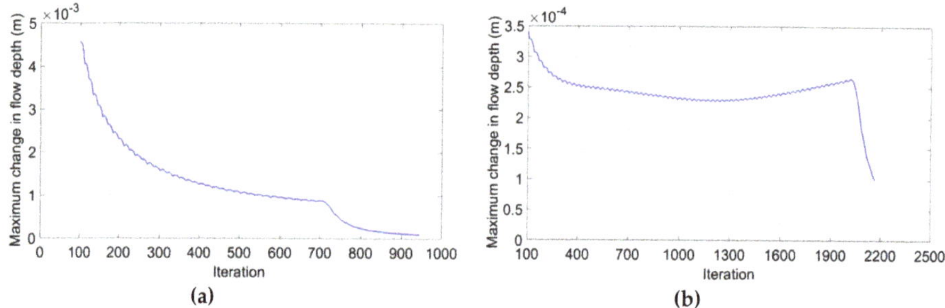

Figure 9. Evolution of the convergence criteria: (**a**) for test case 2 with Dissipative scheme with Boussinesq terms; (**b**) for test case 5 with MacCormack scheme without Boussinesq terms.

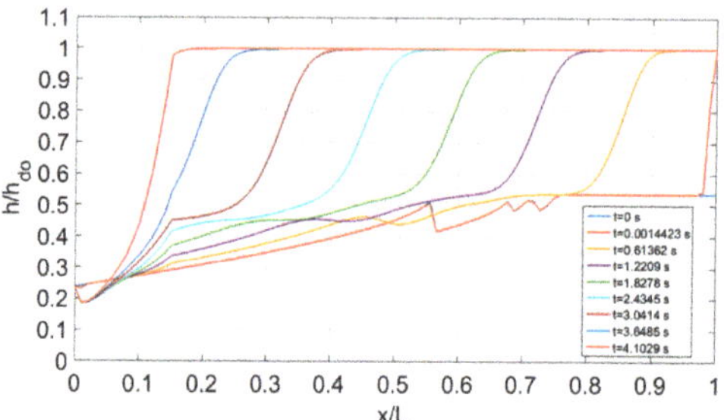

Figure 10. Temporal evolution of jump formation for test case 3 using the Dissipative two-four scheme including the Boussinesq terms.

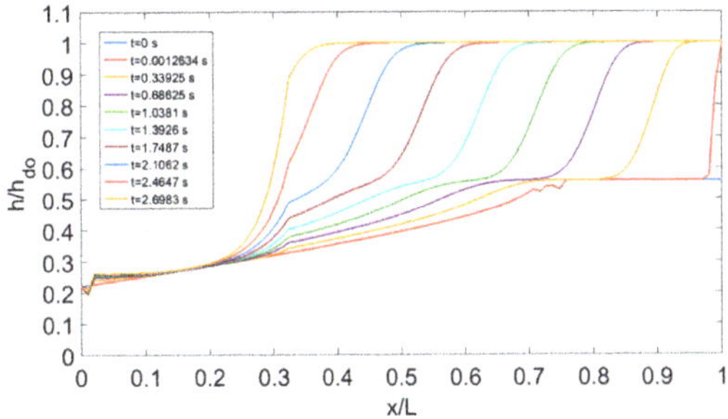

Figure 11. Temporal evolution of jump formation for test case 4 using the Dissipative two-four scheme including the Boussinesq terms.

The specific force of the flow (momentum flux per unit weight) in a channel with rectangular cross section written as $M = 0.5bh^2 + \left(Q^2/gbh\right)$, where b is the width of the rectangular cross section, is conserved across a hydraulic jump, or alternatively, it takes the same value upstream and downstream of the jump. From the numerical results it turns out that the relative error of the specific force between the two end cross-sections of the jump for all test cases and the three numerical schemes applied ranged from 0.62 % to 12.92 %, results that are acceptable for the accurate location of the jump. The number of iterations required to obtain a steady state with accuracy of order 10^{-4} m or 5×10^{-5} m, along with the closure of continuity equation for each test case and all numerical schemes are shown in Table 2, where the flow rate was computed from the depth and the average velocity over the cross section. It is evident that for all test cases the MacCormack algorithm gave the highest error in mass conservation, if compared to the other algorithms, due to the lower order of spatial accuracy.

Table 2. Required iterations for convergence and mass balance error.

Test Case	Numerical Scheme	Iterations	Maximum Mass Conservation Error (%)
1	Dissipative with Boussinesq terms	3099	0.39
	Dissipative without Boussinesq terms	3098	0.39
	MacCormack without Boussinesq terms	3722	2.79
2	Dissipative with Boussinesq terms	946	0.72
	Dissipative without Boussinesq terms	946	0.73
	MacCormack without Boussinesq terms	1243	2.77
3	Dissipative with Boussinesq terms	3375	0.55
	Dissipative without Boussinesq terms	3146	0.55
	MacCormack without Boussinesq terms	3173	2.86
4	Dissipative with Boussinesq terms	2296	0.78
	Dissipative without Boussinesq terms	1807	0.79
	MacCormack without Boussinesq terms	2219	2.81
5	Dissipative with Boussinesq terms	2218	0.93
	Dissipative without Boussinesq terms	2217	0.92
	MacCormack without Boussinesq terms	2164	2.95
6	Dissipative with Boussinesq terms	2874	1.08
	Dissipative without Boussinesq terms	2873	1.09
	MacCormack without Boussinesq terms	2603	3.02

An iterative algorithm was implemented for both numerical schemes which converged to the steady state solution, when the maximum value of the difference of the flow depth between two successive iterations was smaller than a threshold value. The sequent depths as well as the location where the jump forms for given flow and boundary conditions, resulted from the steady state solution of the governing equations. The numerical results showed that the magnitude of the Boussinesq terms is greater at the jump regime, whereas outside this it is almost zero, as expected. It was also shown that the Boussinesq terms do not affect practically the location where the jump forms for test cases 1–6, so that their exclusion eases the discretization without significant error. Note that to obtain the results presented in this paper artificial viscosity has been used. A coefficient adjusting the amount of dissipation to reproduce the experimental measurements by a trial and error procedure was set to the value 0.011.

5. Case Studies

For further validation of the numerical results obtained from the six test cases examined, let us present a case study of practical interest. We consider a 100 m long and 2 m wide (narrow channel, depth to width ratio around one) rectangular, horizontal open channel, with Manning roughness coefficient 0.013 and flow rate 7.2 m^3/s. In Figure 12a–c we present the computed dimensionless flow depth, h/h_{do}, versus the dimensionless distance x/L, for three Froude numbers at the vena contracta cross-section namely 2, 4 and 6, including or excluding the Boussinesq terms, using a dissipation parameter equal to 0.011. In Figure 12d we demonstrate the temporal evolution of the jump until it is stabilized, for Froude number equal to 4. In all three cases one may notice the formation of H3 and H2 types of free surface curves upstream and downstream of the jump respectively, due to energy loss. From Figure 12b,c it is evident that the hydraulic jump is much longer when computed using Boussinesq terms.

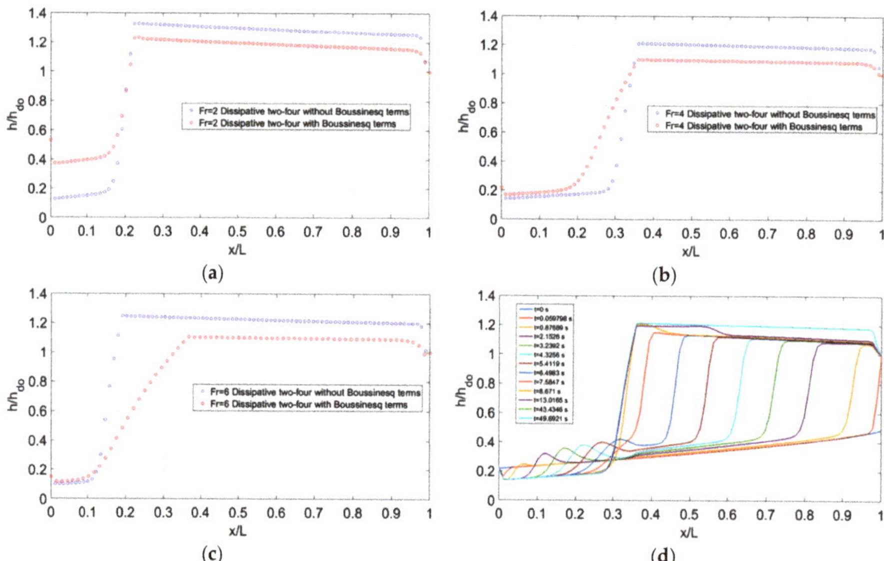

Figure 12. Dimensionless free surface elevation for 2 m wide channel: (**a**) for Fr = 2; (**b**) for Fr = 4; (**c**) for Fr = 6; (**d**) temporal evolution of the jump for Fr = 4.

Let us now examine the influence of the channel width on the length of the hydraulic jump, L_j. We consider a 100 m long channel and 5 m wide (wide section, depth to width ratio smaller than one) with orthogonal cross section conveying 10 m^3/s, with same dissipation factor equal to 0.011. In Figure 13a–c the computed dimensionless flow depth, h/h_{do}, is plotted versus the dimensionless distance x/L, for three Froude numbers at the vena contracta namely 2, 4 and 6, with or without the Boussinesq terms incorporated. The flow depth at x/L = 0 is the boundary condition, and it is the same with and without Boussinesq terms, as indicated in Figures 12 and 13. Near the supercritical boundary the depth deviates from that at x/L = 0, with deviation being greater when Boussinesq terms are not included. This is probably due to numerical instability from the boundary condition, also shown clearly in Figures 2b, 3b, 4b, 5b, 6b and 7b, where Boussinesq terms are plotted vs x/L. The non-Boussinesq terms showed greater instability because the depth near the supercritical boundary deviates more than that with Boussinesq terms.

The length of the hydraulic jump computed with Boussinesq terms, is still greater than that computed without incorporating Boussinesq terms, but the difference is not as sound as it was in the previous example. In Tables 3 and 4 we present the length of the computed hydraulic jump and compare it to that from experiments [27] (p. 14). The results in the 5 m wide channel with or without the Boussinesq terms are in satisfactory agreement with measurements in orthogonal channels, if compared to the results of the 2 m wide channel. The noticeable differences between the flow profiles in Figures 12 and 13 are probably due to the non-hydrostatic pressure distribution encountered in Boussinesq terms. The difference is sound when the hydraulic jump is "weak" for low Froude numbers, where the transition from supercritical to subcritical flow is not as vigorous as for greater Froude numbers. Therefore there is a smoother free surface elevation when Boussinesq terms are included, resulting in longer jump lengths.

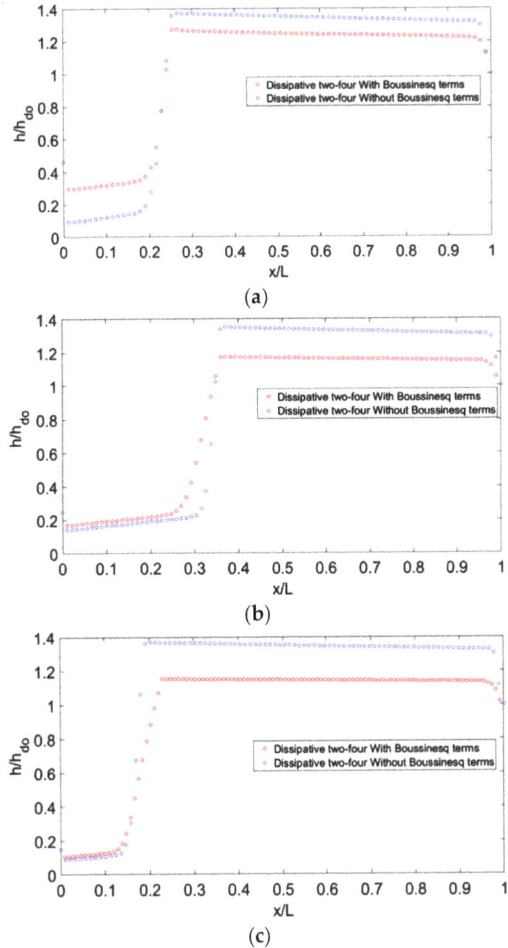

Figure 13. Dimensionless free surface elevation for 5 m wide channel: (**a**) for Fr = 2; (**b**) for Fr = 4; (**c**) for Fr = 6.

Table 3. Length of jump for 2 m wide open channel.

Fr	Scheme	$Lj_{computed}$ (m)	$Lj_{experimental}$ (m)	% Difference
2	Dissipative with Boussinesq terms	10.12	8.81	14.86
	Dissipative without Boussinesq terms	10.12	9.51	6.41
4	Dissipative with Boussinesq terms	15.74	13.14	19.79
	Dissipative without Boussinesq terms	12.36	14.47	−14.58
6	Dissipative with Boussinesq terms	26.47	15.07	75.65
	Dissipative without Boussinesq terms	8.99	17.03	−47.21

From Table 3, it is obvious that comparing the numerical results with experiments, exclusion of Boussinesq terms leads to more accurate results concerning the length of the jump. Alternatively, from Table 4 the opposite is true. Apparently, the flow in a "wide" orthogonal open channel is not affected from the side walls while in a "narrow" channel, it is. Therefore, the one-dimensional shallow water equations that are improved with Boussinesq terms can predict the characteristics of a hydraulic jump accurately, if the channel is wide if compared to the flow depth.

Table 4. Length of jump for 5 m wide open channel.

Fr	Scheme	$Lj_{computed}$ (m)	$Lj_{experimental}$ (m)	% Difference
2	Dissipative with Boussinesq terms	7.59	7.02	8.11
	Dissipative without Boussinesq terms	7.59	7.56	0.40
4	Dissipative with Boussinesq terms	8.99	8.41	6.90
	Dissipative without Boussinesq terms	7.87	9.69	−18.78
6	Dissipative with Boussinesq terms	9.18	10.74	−14.52
	Dissipative without Boussinesq terms	6.74	12.74	−47.10

6. Conclusions

The one-dimensional Boussinesq equations have been solved numerically using the MacCormack as well as the Dissipative two-four finite difference schemes, for the simulation of hydraulic jump formation in a horizontal rectangular open channel and for upstream Froude numbers Fr in the range 2.44 to 5.38. The governing equations have been enriched with additional terms if compared to the Saint Venant equations, to account for the non-hydrostatic pressure distribution in the regime of rapidly varied flow. Terms related to the energy loss and the gravity forces have been also included. The initial condition was a steady supercritical gradually varied flow along the whole length of the channel modeled. The upstream and downstream boundary conditions regarding the flow depth remained constant during the iteration process, and equal to the values measured in experiments. The method of specified intervals was used for the calculation of the velocity at the downstream end, assuming that the positive characteristic through a point does not intersect with already established grid points. Variable time step was used in every iteration according to the CFL (Courant-Friedrichs-Lewy) stability criterion, along with artificial viscosity for smoothing of the oscillations occurring in the jump. The computational results compare well with experiments since the specific force was computed from the depth and mean velocity at both ends of the hydraulic jump with acceptable tolerance, and the mass conservation equation was verified for all numerical schemes and all test cases.

From such a model one can determine the sequent depth ratio as well as the length of the jump, results that are useful in the design of stilling basins (geometrical properties). Given a stilling basin with a known inflow Froude number and flow depth, the engineer must decide the end sill dimensions and the basin length, so that the hydraulic jump is contained in the stilling basin. Finally, from comparison of the numerical results and experiments, it can be concluded that the aforementioned numerical modeling schemes can predict the basic features of the classical hydraulic jump with acceptable accuracy.

Author Contributions: E.R. has performed the experimental measurements and implemented the numerical algorithms. E.R. wrote the paper with the contribution of P.P. All authors have read and agreed to the published version of the manuscript.

Funding: This research received no external funding.

Acknowledgments: The authors would like to thank J. Demetriou for his contribution in the implementation of the experiments.

Conflicts of Interest: The authors declare no conflict of interest.

References

1. Bidone, G. *Observations Sur Le Hauteur Du Ressaut Hydraulique en 1818, Report*; Royal Academy of Sciences: Turin, Italy, 1819. (In French)
2. Belanger, J.B. *Essai Sur La Solution Numeric de Quelques Problems Relatifs an Mouvement Permenent Des Causcourantes*; Carilian-Goeury: Paris, France, 1828. (In French)
3. Bradley, J.N.; Peterka, A.J. Hydraulic Design of Stilling Basins. *J. Hydraul. Div.* **1957**, *83*, 1401–1406.
4. Hager, H.W.; Bremen, R. Classical Hydraulic Jump: Sequent Depths. *J. Hydraul. Res.* **1989**, *27*, 565–583. [CrossRef]
5. Hager, H.W.; Bremen, R.; Kawagoshi, N. Classical Hydraulic Jump: Length of Roller. *J. Hydraul. Res.* **1990**, *28*, 591–608. [CrossRef]
6. Long, D.; Rajaratnam, N.; Steffler, P.; Smy, P. Structure of Flow in Hydraulic Jumps. *J. Hydraul. Res.* **1991**, *29*, 207–218. [CrossRef]
7. Wu, S.; Rajaratnam, N. Free Jumps, Submerged Jumps and Wall Jets. *J. Hydraul. Res.* **1995**, *11*, 197–212. [CrossRef]
8. Katopodes, N.D. A Dissipative Galerkin Scheme for Open-Channel Flow. *J. Hydraul. Eng.* **1984**, *110*, 450–466. [CrossRef]
9. Rotunno, R.; Smolarkiewicz, K.P. Vorticity Generation in the Shallow-Water Equations as Applied to Hydraulic Jumps. *J. Atmos. Sci.* **1995**, *52*, 320–330. [CrossRef]
10. Khan, A.A.; Steffler, M.P. Physically Based Hydraulic Jump Model for Depth-Averaged Computations. *J. Hydraul. Eng.* **1996**, *122*, 540–548. [CrossRef]
11. Su, L.B.; Wei, L.W. Numerical Simulation of 2D Flows with Hydraulic Jump Using Shallow Water Equations. *Appl. Mech. Mater.* **2011**, *130–134*, 3616–3619. [CrossRef]
12. Liu, L.Y.; Cai, X.Y.; Wei, L.W.; Yang, K.; Ma, Z. Numerical Simulation of Hydraulic Jump Using Eno Scheme. *J. Chem. Pharm. Res.* **2014**, *6*, 603–607.
13. Qingchao, L.; Drewes, U. Turbulence Characteristics in Free and Forced Hydraulic Jumps. *J. Hydraul. Res.* **1994**, *32*, 877–898. [CrossRef]
14. Chippada, S.; Ramaswamy, B.; Wheeler, F.M. Numerical Simulation of Hydraulic Jump. *Int. J. Numer. Methods Eng.* **1994**, *37*, 1381–1397. [CrossRef]
15. Castillo, G.L.; Carrillo, M.J.; García Antonio, T.J.; Rodríguez, V.A. Numerical Simulations and Laboratory Measurements in Hydraulic Jumps. In Proceedings of the 11th International Conference on Hydroinformatics HIC 2014, New York, NY, USA, 17–21 August 2014; pp. 1–8.
16. Barrachina, B.A.; Jimenez, A.P. Numerical Analysis of Hydraulic Jumps Using OpenFOAM. *J. Hydroinform.* **2015**, *17*, 662–678. [CrossRef]
17. Mortazavi, M.; Chenadec, V.; Moin, P.; Mani, A. Direct Numerical Simulation of a Turbulent Hydraulic Jump: Turbulence Statistics and Air Entrainment. *J. Fluid Mech.* **2016**, *797*, 60–94. [CrossRef]
18. Valero, D.; Viti, N.; Gualtieri, C. Numerical Simulation of Hydraulic Jumps. Part 1: Experimental Data for Modelling Performance Assessment. *Water* **2019**, *11*, 36. [CrossRef]
19. Viti, N.; Valero, D.; Gualtieri, C. Numerical Simulation of Hydraulic Jumps. Part 2: Recent Results and Future Outlook. *Water* **2019**, *11*, 28. [CrossRef]
20. Basco, D.R. *Computation of Rapidly Varied, Unsteady, Free Surface Flow*; U.S. Geological Survey, Water Resources Investigations Report No. 83-4284; Geological Service: Reston, VT, USA, 1987.
21. MacCormack, R.W. The Effect of Viscosity in Hypervelocity Impact Cratering. In Proceedings of the AIAA Hypervelocity Impact Conference, Cincinnati, OH, USA, 30 April–2 May 1969; pp. 1–7. [CrossRef]

water profile, which can be obtained by performing a time-dependent simulation with steady boundary conditions over a period of time that is long enough to reach a steady solution [14]. This step may consume a considerable amount of time before the main problem can be addressed. The initial condition should be computed with the same numerical scheme as the one used in the unsteady model in order to ensure the steadiness of the first step of the unsteady model. The ability of these models to quickly obtain a steady initial solution is also of great importance.

Optimization is another field where obtaining a steady result as quickly as possible is important. Indeed, most optimization techniques require a large number of runs in order to figure out the optimal solution. In order to ensure that the overall computation time is as short as possible, techniques that are quick should be utilized including parallelization [15] or the use of fast computing models.

Although 1-D simulations are known to provide results in a short period of time, accelerating the computation of the 1-D steady solution is of great interest in the fields mentioned above. This can be achieved by using two main strategies. The first consists of exploiting the resources of modern computers more efficiently. Such techniques are more frequently applied to 2-D cases, which naturally require more computing resources. Common hardware acceleration strategies include parallelizing codes on several CPU cores [16,17] or on a GPU [18–20]. The second method consists of designing algorithms in order to converge with less effort toward the solution. To the authors' knowledge, there is no such work available in the literature. Both strategies can be combined in order to obtain the best performance.

The purpose of this paper is to propose algorithmic strategies for the fast computation of 1-D steady solutions. First, the equations are presented. Then, we introduce two original strategies in order to reduce the overall computation time. These strategies can easily be implemented in other hydraulic codes. An alternative non-linear solver is introduced for comparison purposes. Finally, the validation results, the optimal parameters for the algorithm, the performance of the model and its application to a real-world problem are presented.

2. Materials and Methods

2.1. Equations

One-dimensional water flow is described by the St-Venant equations, which are as follows [21]:

$$\begin{array}{l} \frac{\partial A}{\partial t} + \frac{\partial Q}{\partial x} = q_l \\ \frac{\partial Q}{\partial t} + \frac{\partial}{\partial x}\left(\frac{Q^2}{A}\right) + gA\frac{\partial z_s}{\partial x} = -gAS_f + u_x q_l \end{array} \tag{1}$$

where A is the cross-section area (m^2), Q is the discharge (m^3/s), q_l is a lateral discharge per unit length (m^2/s), g is the gravity acceleration (m/s^2), S_f is the friction slope (-), u_x is the velocity along the x direction of q_l (m/s) and z_s is the free surface elevation (m). The numerical resolution of this set of non-conservative equations has been shown to provide accurate results in many practical cases, including in the presence of discontinuities [22,23].

Assuming a steady flow, temporal derivatives vanish in Equation (1). Since solving the first equation is straightforward, the discharge distribution is known on the entire domain for a given distribution of q_l. We assume that the sign of Q is independent from x. It means that a single equation remains with a single unknown A. In order to keep the same numerical scheme as the one used for the unsteady system (which is important when a steady solution is used as the initial condition of an unsteady problem and to be able to use a similar algorithmic strategy), Kerger et al. [14] added a pseudo-temporal term (pseudo-time is τ (s)) to the steady form of Equation (1):

$$\beta\frac{\partial A}{\partial \tau} + \frac{\partial}{\partial x}\left(\frac{Q^2}{A}\right) + gA\frac{\partial z_s}{\partial x} = -gAS_f + u_x q_l \tag{2}$$

where $\beta = -\text{sign}(Q)$. Kerger et al. [14] justify this choice for β by analyzing the characteristic velocity of Equation (2):

$$\lambda = \frac{c^2 - u^2}{\beta} = \frac{c^2\left(1 - \text{Fr}^2\right)}{\beta} \tag{3}$$

where c (m/s) is the wave celerity. In subcritical flows (Fr < 1, $\text{Fr} = Q/\left(gA^3/b\right)^{0.5}$ is the Froude number (-), b is the width of the free surface (m)), and the sign of λ is the sign of β. If Fr > 1, then $\text{sign}(\lambda) = -\text{sign}(\beta)$. For critical flows (Fr = 1), the characteristic velocity is 0, independently from β. In order to keep some form of consistency with the general model, if we choose $\beta = -\text{sign}(Q)$, an upstream boundary condition is required when Fr > 1 and a downstream boundary condition is required when Fr < 1. This is equivalent to the position of the water depth boundary condition for the numerical resolution of the full 1-D set of equations.

With a single boundary condition and a discharge distributed in the channel, solving Equation (2) determines the cross-section area (and subsequently, the water depth) all along the stretch. Equation (2) is discretized according to the finite volume method. It is solved according to the same numerical scheme as the one used for the full unsteady model [14,24].

For a node i, Equation (2) is discretized in finite volumes as:

$$\frac{Q_{i+1/2}^2/A_{i+1/2} - Q_{i-1/2}^2/A_{i-1/2}}{\Delta x} + gA_i\left(\frac{z_{s,i+1/2} - z_{s,i-1/2}}{\Delta x} + S_{f,i}\right) - u_{x,i}q_{l,i} \approx -\beta\frac{\partial A}{\partial \tau} \tag{4}$$

where Δx (m) is the spatial discretization step and subscripts refer to the position of variables values.

For the sake of clarity, we consider $Q > 0$ and a constant reconstruction of the flux at finite volume boundaries to explicate the numerical scheme. When applying the considered upwinding directions [14] for a node i not located next to a boundary, Equation (4) is equivalent to:

$$\frac{Q_i^2/A_i - Q_{i-1}^2/A_{i-1}}{\Delta x} + gA_i\left(\frac{z_{s,i+1} - z_{s,i}}{\Delta x} + S_{f,i}\right) - u_{x,i}q_{l,i} = \frac{\partial A}{\partial \tau} \tag{5}$$

This flux vector splitting method has been shown to be unconditionally stable [14]. For the node located at the downstream boundary ($i = N-1$), if a weak water level boundary condition $z_{s,BC}$ is imposed at the border, Equation (4) becomes:

$$\frac{Q_{N-1}^2/A_{N-1} - Q_{N-2}^2/A_{N-2}}{\Delta x} + gA_{N-1}\left(\frac{z_{s,BC} - z_{s,N-1}}{\Delta x} + S_{f,N-1}\right) - u_{x,N-1}q_{l,N-1} = \frac{\partial A}{\partial \tau} \tag{6}$$

Without a boundary condition imposed on the value of z_s at the external border, a nil z_s gradient is imposed and Equation (4) becomes:

$$\frac{Q_{N-1}^2/A_{N-1} - Q_{N-2}^2/A_{N-2}}{\Delta x} + gA_{N-1}S_{f,N-1} - u_{x,N-1}q_{l,N-1} = \frac{\partial A}{\partial \tau} \tag{7}$$

At the upstream node ($i = 0$), if a weak boundary condition of the water level is imposed at the border, Equation (4) becomes:

$$gA_0\left(\frac{z_{s,1} - z_{s,BC}}{\Delta x} + S_{f,0}\right) - u_{x,0}q_{l,0} = \frac{\partial A}{\partial \tau} \tag{8}$$

Without a boundary condition at the upstream border, Equation (4) becomes:

$$gA_0\left(\frac{z_{s,1} - z_{s,0}}{\Delta x} + S_{f,0}\right) - u_{x,0}q_{l,0} = \frac{\partial A}{\partial \tau} \tag{9}$$

2.2. *Original Solving Strategy*

Solving Equation (2) instead of Equation (1) decreases the computation time since the number of equations and unknowns is reduced. In order to save even more time, two additional strategies were implemented: (a) a non-linear Krylov accelerator was used to promote fast convergence and (b) the computation was only performed on a sliding part of the full domain.

2.2.1. Non-Linear Krylov Acceleration

Numerically solving a non-linear system can be performed by different means, including Newton's method and Broyden's method [25]. More sophisticated methods exist in order to solve non-linear systems faster, such as the Jacobian-free Newton–Krylov method [26] and Anderson acceleration [27]. The Anderson acceleration method uses the results from successive iterations in order to adapt the new approximation. Walker and Ni [28] showed that this method can be considered equivalent to the well-known GMRES method [29] when applied to linear systems. The nonlinear Krylov acceleration (NKA) [30,31], which is similar to Anderson acceleration, has been found to be more efficient in some applications than more recent methods such as the Jacobian-free Newton–Krylov method [32]. NKA was used for a faster convergence of our pseudo-time model.

Since NKA only relies on the results directly produced by the hydraulic model, it can be easily applied to other algorithms or other domains. Indeed, no gradient evaluation (nor other prerequisite) is required before calling on the NKA algorithm.

The NKA algorithm records N ($N \in \mathbb{N}_{>0}$) previous moves of the root finding process. Based on these previous moves, NKA adapts its guess for the new root. One of the main assumptions is that the Jacobian matrix remains constant within the scope of N moves. We briefly explain the method here and extended details can be found in [30–33].

A Newton–Raphson iteration process computes the n + 1st guess of the root based on the nth guess $\mathbf{x}_n$, on the value of the function f at $\mathbf{x}_n$ and on the invert of the Jacobian matrix J:

$$\mathbf{x}_{n+1} = \mathbf{x}_n - \mathrm{J}^{-1} f(\mathbf{x}_n) \tag{10}$$

Instead of evaluating J^{-1} at each iteration, NKA evaluates it from N previous moves or sets it to the identity matrix, in which case the method degenerates to the fixed-point method.

NKA takes advantage of a history of N corrections of $\mathbf{x}$ (denoted $\mathbf{v}$) and N evolutions of $f(\mathbf{x})$ (denoted $\mathbf{w}$) at iterate n:

$$\begin{array}{l} \mathbf{v}_i = \mathbf{x}_i - \mathbf{x}_{i-1} \\ \mathbf{w}_i = f(\mathbf{x}_i) - f(\mathbf{x}_{i-1}) \end{array}, i = n - N + 1, \ldots, n \tag{11}$$

The method assumes that J is constant and invertible within the scope of the N previous iterations, which is written as follows:

$$\begin{array}{l} \mathrm{J}\mathbf{v}_i = \mathbf{w}_i \\ \mathbf{v}_i = \mathrm{J}^{-1}\mathbf{w}_i \end{array} \tag{12}$$

Mathematical developments described in the references cited above lead to the expression:

$$\begin{array}{l} \mathbf{v}_{n+1} = \sum\limits_{i=n-N+1}^{n} z_i \mathbf{v}_i + \left(f(\mathbf{x}_n) - \sum\limits_{i=n-N+1}^{n} z_i \mathbf{w}_i \right) \\ \mathbf{x}_{n+1} = \mathbf{x}_n + \mathbf{v}_{n+1} \end{array} \tag{13}$$

where the coefficients z_i are the solution of the projection:

$$\mathbf{z} = \mathrm{argmin}_{a \in \mathbb{R}_0} \left\| f(\mathbf{x}_n) - \sum_{i=n-N+1}^{n} a_i \mathbf{w}_i \right\| \tag{14}$$

Equation (13) shows that the correction of the variable $\mathbf{x}$ is decomposed into two components:

1. The first term depicts the correction as a linear combination of previous corrections.
2. The second term is similar to the second term of a fixed-point iteration that takes into account previous evolutions of function *f*.

Note that the Jacobian matrix is not used in this iterative process.

2.2.2. Evolutionary Domain

The second strategy was designed to reduce computational cost. It consists in reducing the size of the computation domain and sliding it along the river stretch in order to evaluate the cross-section areas from downstream to upstream.

The development of this strategy has arisen from the long history of numerical hydrodynamics in various flow regimes. Indeed, many flows that are solved for rivers are subcritical (Fr < 1) at downstream and upstream boundary conditions. In such a situation, the cross-section area information is propagated from downstream to upstream. For a fixed flow direction, the upwinding direction of the scheme takes all unknowns and properties (except those linked to the discharge) downstream. This means that when a new node is computed, it depends only on the downstream nodes (Figure 1). It should be noticed that when a node i is added, the upstream border of node $i+1$ produces a change in the upwinding direction of property Q and the unknown A_{vel} (cross-section area used to compute the velocity). The node $i+1$, which was supposed to be converged, undergoes a new convergence process that indirectly affects nodes $i+2$, $i+3$, … Theoretically, all the nodes located downstream of node i should be kept in the computational domain.

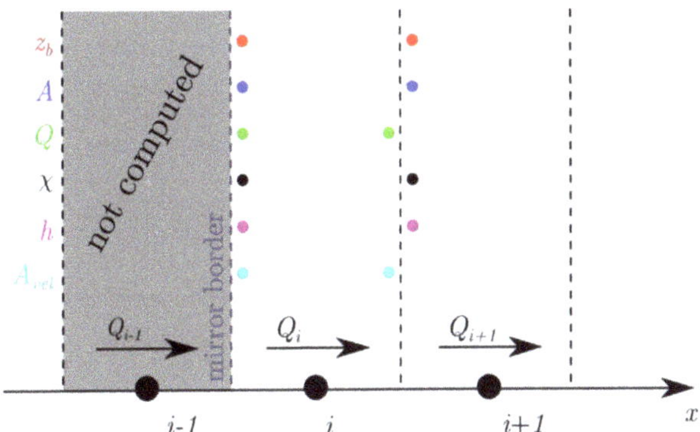

Figure 1. Upwinding directions of the unknown and flow properties at the upstream limit of the computation domain.

The boundary condition on the border between the node i and node $i-1$ (see Figure 1) is called a "mirror border". On this border, all the unknowns and properties are reconstructed from the computed inner node. This method is equivalent to reproducing the node i in $i-1$, like a mirror. For node i in Figure 1, the discretization of Equation (2) is:

$$gA_i\left(\frac{z_{s,i+1} - z_{s,i}}{\Delta x} + S_{f,i}\right) - u_{x,i}q_{l,i} = 0 \tag{15}$$

The evolution of the domain relies on three strategies. First, all nodes in the computation domain should have reached a partial residual threshold ξ_p (m^2/s) before considering the extension of the computation domain: $\partial A/\partial \tau \leq \xi_p$. Then, the most downstream node can be removed from the partial domain once it drops under the final residual threshold ξ_f (m^2/s): $\partial A/\partial \tau \leq \xi_f$. Finally, we have to make sure that the nodes that were removed from the domain are not impacted by the computation domain in such a way that their residuals $\partial A/\partial\tau$ become higher than the final precision threshold ξ_f. Dimensionless parameters are discussed further in Section 3.2.

In order to assess whether the removed nodes $\partial A/\partial\tau$ remain lower than ξ_f, an analytical analysis of the discretized model is performed. Let us consider three nodes and their borders (Figure 2). The discretized formulation of Equation (2) at node i is:

$$\begin{aligned}\left[\frac{\partial A}{\partial\tau}\right]_i &= \left[\frac{\partial}{\partial x}\left(\frac{Q^2}{A}\right)+gA\frac{\partial z_s}{\partial x}+gAS_f-u_xq_l\right]_i \\ &\approx \frac{\left(\frac{Q_i^2}{A_i}-\frac{Q_{i-1}^2}{A_{i-1}}\right)}{\Delta x}+g\frac{A_{i+1}^2-A_i^2}{A_{i+1}-A_i}\left(\frac{z_{s,i+1}-z_{s,i}}{\Delta x}\right)+\left[gAS_f-u_xq_l\right]_i\end{aligned} \tag{16}$$

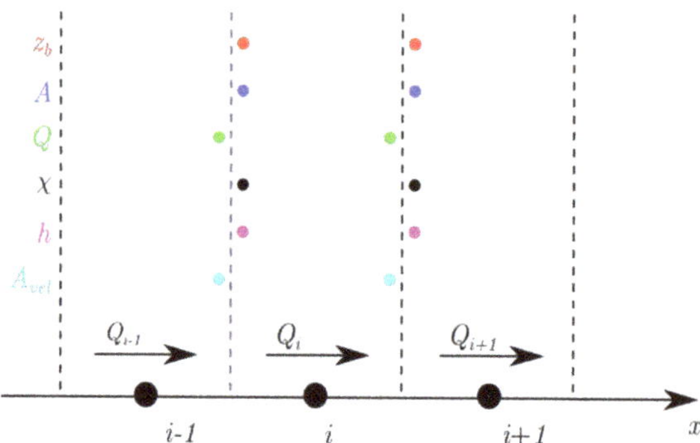

Figure 2. Upwinding directions of the unknown and flow properties in the computation domain.

The influence of a change in the cross-section area in $i-1$ on the value of $\partial A/\partial\tau$ in i is given by the derivative:

$$\frac{\partial}{\partial A_{i-1}}\left(\left[\frac{\partial A}{\partial\tau}\right]_i\right)=\frac{Q_{i-1}^2}{\Delta x}\frac{1}{A_{i-1}^2} \tag{17}$$

From the result in (17), one can predict the evolution of the residual $\partial A/\partial\tau$ of a node that is not in the computation domain anymore. If node i is planned to be removed from the computation domain, one can check that its $\partial A/\partial\tau$ remains below the threshold ξ_f even with an evolution of the cross-section area in $i-1$:

$$\left[\frac{\partial A}{\partial\tau}\right]_{i,next}\approx\left[\frac{\partial A}{\partial\tau}\right]_{i,prev}+\frac{Q_{i-1}^2}{\Delta x}\frac{1}{A_{i-1}^2}\Delta A_{i-1} \tag{18}$$

where ΔA_{i-1} is the evolution of the cross-section A in $i-1$ from the last evaluation of $[\partial A/\partial\tau]_{i,prev}$.

If the evolution of the cross-section area in the computation domain ($i-1$) implies that $[\partial A/\partial\tau]_i$ exceeds the threshold ξ_f, then node i should be added again in the computation domain in order to make sure it remains below the threshold until the end of the entire computation. It should be noted that node $i+1, i+2, \ldots$ might also be impacted. This technique guarantees that once the sliding domain has moved along the entire domain, all nodes have reached at least the final precision required.

As stated earlier, the method described here was designed within the framework of subcritical flows. In order to be able to deal with a larger range of flow regimes, several adaptations were made. When a supercritical node is detected downstream (let say at position m), it is not computed and the computation domain is extended until a subcritical node is found upstream (say at position n). Then, the domain starting from m to n is computed and converged. This technique avoids boundary condition problems. Indeed, a supercritical flow requires an upstream BC since the characteristic velocity is directed towards the downstream.

2.2.3. Combination of Solving Strategies

The combined use of the sliding domain and NKA involves several specific considerations. The use of NKA is implemented in the code with a safety coefficient that deactivates this optimizing technique in some cases. Indeed, it was experienced that NKA could lead to some instabilities when there was a sudden change in the cross-section area. This behavior is due to the assumption in NKA that the Jacobian matrix is constant and invertible locally [32,33]. In order to avoid such a situation, the accelerator is deactivated when $|A_{i+1} - A_i| > \eta A_{i+1}$, where A_i and A_{i+1} refer to the cross-section area of the nodes i and i + 1 as depicted in Figure 1, and $0 < \eta \leq 1$ is the safety coefficient (-). After several tests, we found that $\eta = 0.5$ provides stability with a limited impact on the computation time.

The method can be summarized with the following pseudo-code (Algorithm 1):

Algorithm 1

```
Initialize (computation list is empty)
Add most downstream node to computation list
While some nodes still have to be converged (∂A/∂τ > ξ_f):
  Initialize lastly added upstream node
  While nodes of computation list not converged (∂A/∂τ > ξ_p):
    Compute cross section change for each node
    If NKA activated:
      Adapt cross section change with NKA algorithm
  If lastly removed downstream node significantly impacted:
    Add it back to the computation list and do not expand upstream
  Else:
    If downstream node in computation list fully converged (∂A/∂τ ≤ ξ_f):
      Remove this node from computation list
  If upstream nodes remain to be added to computation list:
    Add 1 upstream node to the computation list
Finalize
```

2.3. *Alternative Non-Linear Solver*

Various techniques can be used to solve nonlinear Equation (2). Up to this point, we have chosen to discretize the equation using a finite volume scheme and solve it with an explicit time scheme, which is consistent with the unified strategy of WOLF [24]. Other techniques can be used, including finite difference schemes and/or implicit time schemes. Another possibility is to use an optimization algorithm for nonlinear systems. One of these is the recently developed CasADi software [34,35].

CasADi first started as an algorithmic differentiation tool. During its evolution, developers chose to shift the focus toward optimization. From non-linear expressions, CasADi is able to generate all the information needed by a nonlinear solver in order to return a solution to the problem. CasADi provides interfaces to MATLAB or Python for easy use.

The purpose of using CasADi is to show how our algorithm performs compared to a state-of-the-art solver.

The implementation in CasADi was done through Opti stack, a collection of helper functions used for nonlinear programming problems. It is possible to define variables to optimize, parameters, an objective function and constraints. The solving of a 1-D steady open channel flow can be done thanks to this framework.

The constraints of the problem are discretized in Equation (2) for each node and a water depth above 0 everywhere. The downstream boundary condition is imposed through Equation (6). If no boundary condition is set, then a flow condition can be imposed through a constraint on the Froude number for the downstream node and the flow head is minimized at the upstream node. If the flow presents a critical section, minimizing the head upstream is equivalent to finding the section with the highest critical head.

Another way to solve a flow with a critical section is to prescribe a Froude number transition fromFr < 1 to Fr > 1 at that critical section. This is done by setting a constraint on the Froude number on the nodes upstream and downstream of the critical section. The identification of the critical section should be done prior to the computation on the basis of a critical head analysis.

3. Results and Discussion

This section presents the validation of the results and focuses on the optimal parameters and performance. The geometries of the tests were different in order to examine as many cases as possible.

3.1. Validation

The validation of the models was performed on a bump placed in a straight horizontal channel, considering three different flow conditions. The bump and channel geometry have been described previously in [36]. The whole domain ranges from 0 to 20 m with the following bed elevation:

$$z_b(x) = \begin{cases} 0.8\left(1 - \frac{(x-10)^2}{4}\right) 8 \text{ m} \leq x \leq 12 \text{ m} \\ 0 \end{cases} \tag{19}$$

The channel is considered to be rectangular. The discretization step was chosen as 0.1 m.

The different flow conditions are described in Table 1. All tests were performed with $\xi_p = 10^{-6}$ m^2/s and $\xi_f = 10^{-10}$ m^2/s.

Table 1. Boundary conditions for three test cases.

Test	Upstream BC	Downstream BC
A	$q = 1$ m^2/s	$h = 1.7$ m
B	$q = 0.4$ m^2/s	$h = 0.75$ m
C	$q = 0.4$ m^2/s	Transmissive

The objective was to show that the model is able to deal accurately with various flow regimes and transitions. Test A simulates a fully subcritical flow with no transition. Test B creates a subcritical flow upstream, a subcritical flow downstream and a hydraulic jump in between, downstream of the bump. Finally, the goal of test C is to show the robustness of the method for a downstream supercritical flow and an upstream subcritical flow.

The analytical solutions for tests A, B and C were computed from the Bernoulli principle (head conservation) [22] and the conjugate water depth formula was used for test B. A graphical comparison of the analytical values and numerical results obtained by our algorithm and CasADi is given in Figures 3–5. It appears that the new algorithm and CasADi provide results that fit well with the analytical solution. However, a small difference in energy can be noticed between the analytical solution and the numerical results and also between both numerical methods (see Table 2). This was quantified and explained by Bruwier et al. [22]. Moreover, a more noticeable localized difference appears between CasADi and the new algorithm results at the hydraulic jump. Even if the numerical scheme is the same for the new algorithm and CasADi, each method has its own convergence threshold. Altogether, the analysis validates both models.

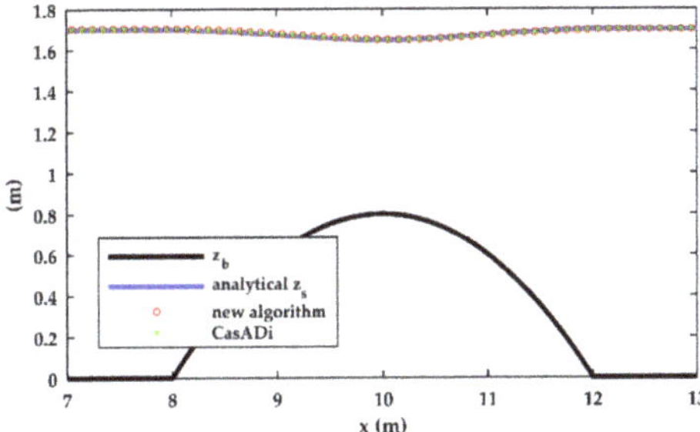

Figure 3. Comparison between the analytical solution and the numerical results produced by the new algorithm and CasADi for validation test A.

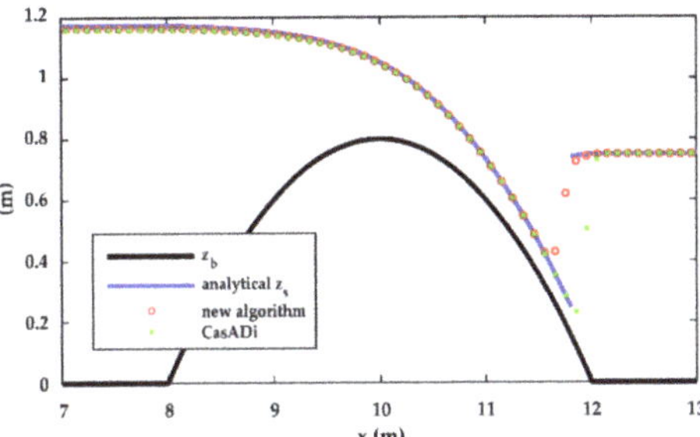

Figure 4. Comparison between the analytical solution and the numerical results produced by the new algorithm and CasADi for validation test B.

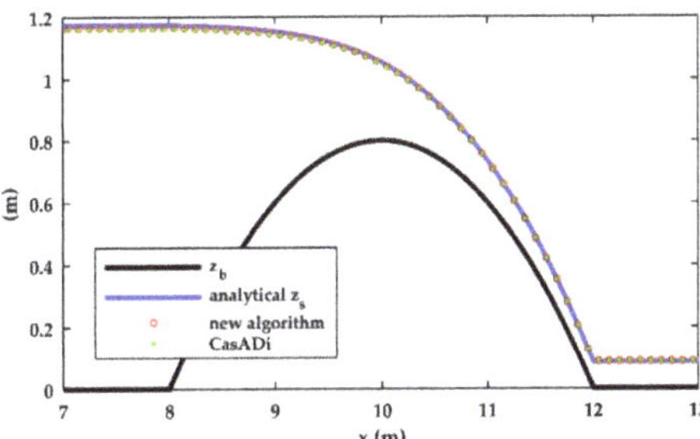

Figure 5. Comparison between the analytical solution and the numerical results produced by the new algorithm and CasADi for validation test C.

Table 2. Upstream head values for tests A, B and C and differential to the analytical value.

Test	Analytical	New Algorithm		CasADi	
	Head (m)	Head (m)	Diff.	Head (m)	Diff.
A	1.71764	1.72147	0.22%	1.71958	0.11%
B	1.18040	1.17062	−0.83%	1.16664	−1.17%
C	1.18040	1.17062	−0.83%	1.16664	−1.17%

3.2. Optimal Setting of the New Algorithm

Five test cases were defined in order to specify the optimal values for the parameters for the new algorithm. These five tests were designed in order to induce changes in the flow characteristics due to topographic or cross-section variations. The first three tests (1 to 3) concern a channel with a rectangular cross-section and a bed slope that follows a sine function:

$$z(x) = \alpha \sin(\beta \pi x) + \gamma \tag{20}$$

with $x \in [0; 20]$, $\alpha = 0.05$, $\gamma = 0.05$, $\beta = 1/2$ (for tests 1 and 2) and $\beta = 2$ for test 3. Two hundred nodes were used to discretize the 20 m long channel, resulting in a 10 cm spatial step. For test 1, the downstream boundary condition is a free surface elevation imposed at 1.2 m. For tests 2 and 3, the same type of boundary condition was imposed with a smaller value of 0.55 m, which results in a higher Froude number downstream. The specific discharge was imposed upstream at 1 m^2/s for tests 1 to 3. The Manning equation was used for friction in tests 1 to 3, with the Manning coefficient n = 0.04 $s/m^{1/3}$.

The topography and the hydraulic solutions were found thanks to the principle of head conservation (Bernoulli) and are shown in Figures 6–8 for tests 1 to 3. Table 3 summarizes the characteristics of each test. The objective of tests 1 to 3 was to analyze the influence of a variation of the bed topography on the behavior of the sliding domain. For test 1, the irregularity of the bed has only a slight influence on the water level. For the other tests, the higher Froude number and less spaced bed elevation variations were meant to investigate the possible influence of oscillations in the water level on the sliding domain performance.

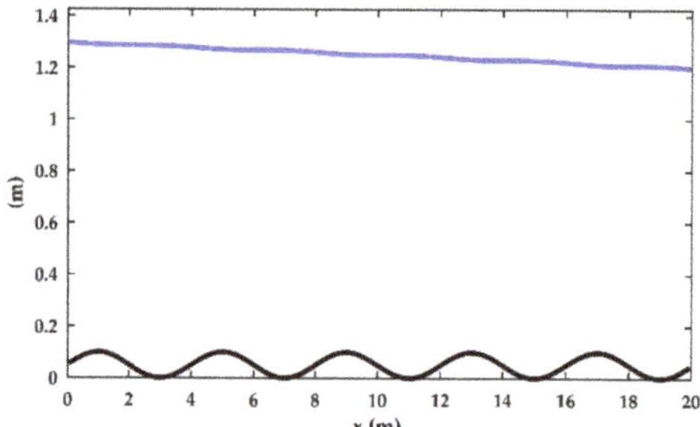

Figure 6. Hydraulic solution for test 1.

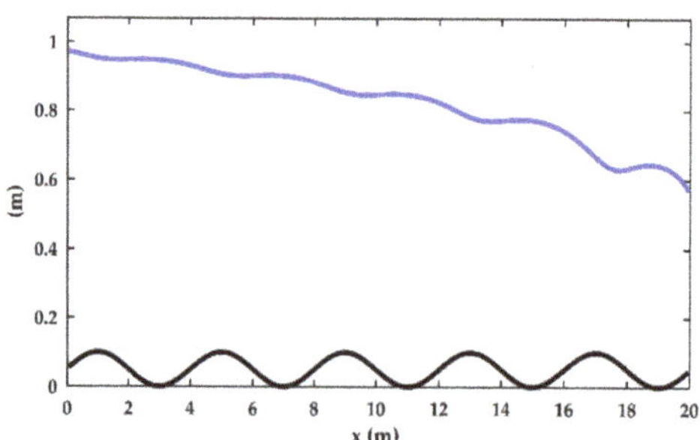

Figure 7. Hydraulic solution for test 2.

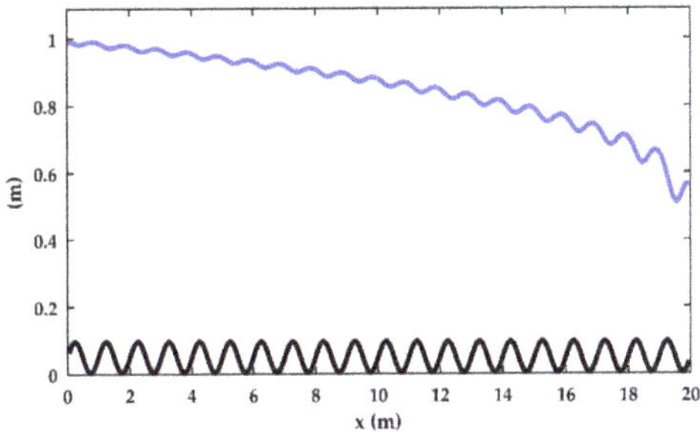

Figure 8. Hydraulic solution for test 3.

Table 3. Summary of the 5 tests used to investigate the best parameters for the new algorithm.

Test	Channel Bed	Friction	BC
1	$z(x) = 0.05\sin(\pi x/2) + 0.05$, rectangular cross-section	Manning, $n = 0.04$ s/m$^{1/3}$	$q = 1$ m/s^2, $z_{down} = 1.2$ m
2	$z(x) = 0.05\sin(\pi x/2) + 0.05$, rectangular cross-section	Manning, $n = 0.04$ s/m$^{1/3}$	$q = 1$ m/s^2, $z_{down} = 0.55$ m
3	$z(x) = 0.05\sin(2\pi x) + 0.05$, rectangular cross-section	Manning, $n = 0.04$ s/m$^{1/3}$	$q = 1$ m/s^2, $z_{down} = 0.55$ m
4	Flat bottom with a weir, rectangular cross-section	Manning, $n = 0.04$ s/m$^{1/3}$	$q = 1$ m/s^2, $z_{down} = 0.7$ m
5	0.2% slope, sudden change in cross-section	Manning, $n = 0.04$ s/m$^{1/3}$	$q = 1$ m/s^2, $z_{down} = 0.7$ m

Tests 4 and 5 were performed on regular bottoms. The discontinuities that we wanted to explore here are linked to a change in the flow regime due to the presence of a weir (test 4) or a severe change in the cross-section (test 5). For test 4, the topography was set at $z = 0$ m for all nodes except for three of them: $z = 0.5$ m at $x = 9.85$ m and $x = 10.05$ m, and $z = 1$ m at $x = 9.95$ m. Friction was computed with the Manning formula and a coefficient $n = 0.04$ s/m$^{1/3}$. The cross-section is uniform along the channel and is rectangular with a width of 1 m. A discharge of 1 m^2/s was injected upstream and a water depth equal to 0.7 m was imposed downstream.

Test 5 deals with a severe change in the cross-section on an inclined bottom. The channel extends 100 m, discretized with 200 nodes. The slope is 0.2%. The cross-section is trapezoidal upstream ($x \leq 47.5$ m), then suddenly becomes rectangular in the middle of the channel (47.5 m $< x <$ 52.5 m), and finally returns to a trapezoidal shape in the downstream part ($x \geq 52.5$ m). The trapezoidal sections have a width at the bottom of 2 m and the banks are inclined with an angle of 45°. The rectangular cross-sections are 1 m wide. The friction and boundary conditions are the same as in test 4.

The topography and final water levels for tests 4 and 5 are depicted in Figures 9 and 10. A summary of these tests is given in Table 3.

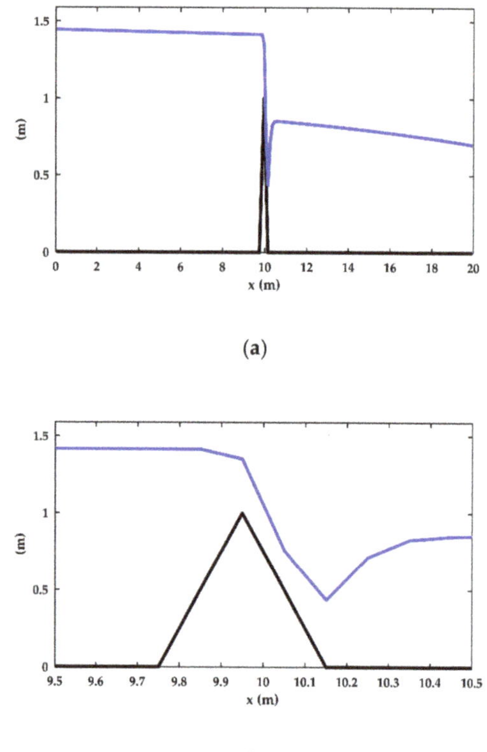

Figure 9. Hydraulic solution for test 4 on the entire domain (**a**) and zoomed on the weir (**b**).

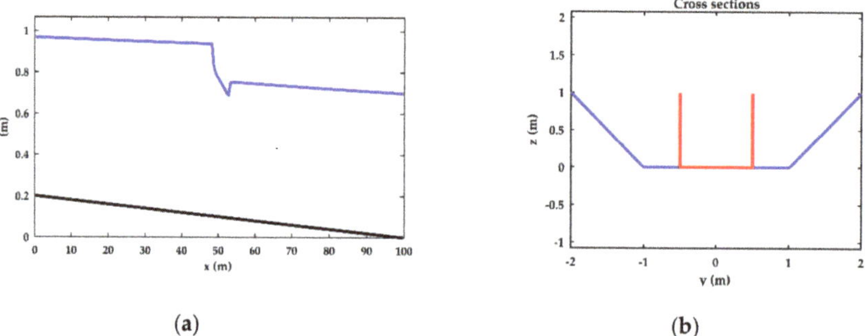

Figure 10. Hydraulic solution (**a**) and cross-section change (**b**) for test 5.

Our algorithm includes many parameters that need to be specified. These parameters are the partial residual threshold ξ_p, the final residual threshold ξ_f, the temporal scheme to solve Equation (4) and the coefficient for the deactivation of the nonlinear Krylov accelerator. The final and partial residuals are two closely linked parameters. They also have a direct impact on the computation time. For a given final residual, which has to be parametrized by the user, the partial residual influences the number of iterates required to converge the partial domain and the size of these domains. After investigation, the other two parameters were shown to have almost no influence on the computation time. The following results focus on the best value to use for the partial residual threshold ξ_p for a fixed value of $\xi_f = 10^{-8}$ m^2/s.

CPU times were measured on a desktop computer (Intel i7 3.5 GHz CPU) for the five tests and various partial residual thresholds. The results are reported graphically in Figure 11. It appears that the overall computation time decreases with an increase in the partial residual threshold. Some stagnation appears around 10^{-2} m^2/s for tests 3 and 5. This can be explained by the fact that the residual naturally decreases at each iteration. Keeping some nodes in the computation domain results in a decrease in the residual for each node included in the computation domain.

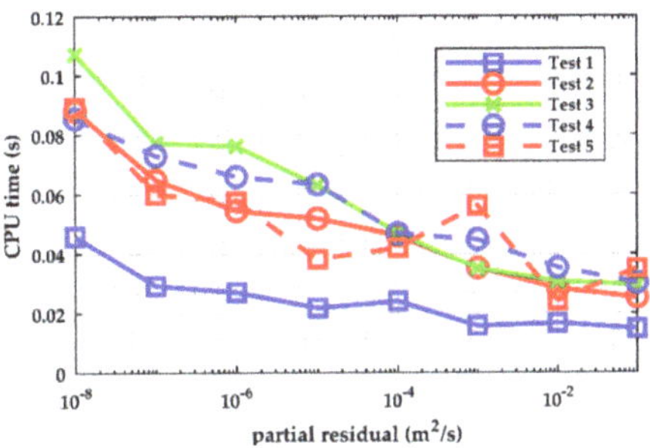

Figure 11. CPU time according to the partial residual threshold for tests 1 to 5.

In order to assess the efficiency of the new algorithm, two scalability tests were performed. The first one consisted of extending the domain upstream, with a constant spatial discretization. The second test consisted of keeping the same channel but refining the discretization and then increasing the number of computation nodes.

The first test took place on a frictionless sine bed elevation described by:

$$z(x) = 0.05\ \sin(2\pi x) + 0.05 \tag{21}$$

The downstream boundary condition is a water level imposed at 0.65 m. The discharge is constant along the channel stretch and is equal to 1 m^2/s. Cross-sections are rectangular and 1 m wide. Four domain lengths were tested (20 m, 200 m, 2 km and 20 km) with a spatial step of 0.1 m, meaning that these domains include 200, 2000, 20,000 and 200,000 nodes.

The computation results (Figure 12a) showed that the computation time per node decreases when the number of nodes increase. This can be explained by the fact that when the domain gets longer, the flow conditions upstream are smoother than downstream. Longer domains undergo fewer changes than shorter domains, leading to shorter computation time per node. This example shows that higher partial residual values provide the best computation times.

In order to complete this scalability study, we looked at the behavior of the algorithm when the spatial step decreases for a given domain length. In classical explicit schemes, this case leads to a quadratic increase in the computation time. Indeed, when the spatial step decreases, the number of nodes increases and the time step decreases.

A 100 km-long channel with a constant 0.025% slope was chosen to illustrate the behavior of the new algorithm. The cross-sections are trapezoidal and are described using tabular values (1 m wide at the bottom of the section and 5 m wide at 1 m above the bottom). Friction was generated using a Manning law with n = 0.03 s/m$^{1/3}$. The downstream boundary condition is a water level set at 1 m, and 1 m^3/s is injected at the upper node and the injection of 4 m^3/s is shared amongst the other nodes (through the q_l term in Equation (1), $u_x = 0$).

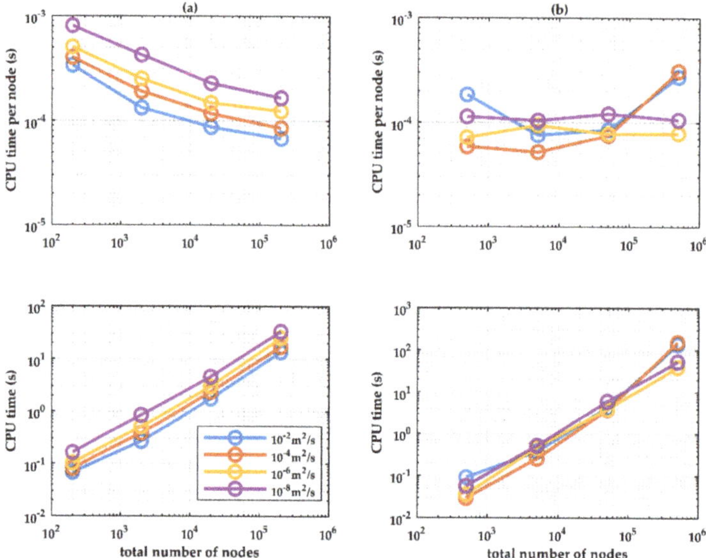

Figure 12. CPU time evolution with the number of nodes when (**a**) the spatial step is kept constant and when (**b**) the domain span is fixed.

The computation times are showed in Figure 12b. The behavior is slightly different from that observed in the previous test. Indeed, the evolution of the computation time is linear in regard to the number of nodes (the CPU time per node is globally constant) when the partial residual is set at 10^{-6} and 10^{-8} m^2/s. For partial residual values of 10^{-2} and 10^{-4} m^2/s, the evolution is linear up to 50,000 nodes; however, for the finest discretization, the computation time increases in a nonlinear way.

This point was investigated further. It appears that at some moment in the computation, the algorithm needs to increase its computation domain size without being able to reduce it quickly (i.e., upstream nodes are added to the computation list while downstream nodes cannot be removed for residual values reasons). This increase in the number of nodes in the computation list was nonlinear compared to the situation with coarser discretization.

We looked at dimensionless values for parameters ξ_p and ξ_f by dividing them by $\sqrt{g}A_0^{3/4}$, A_0, which is a characteristic cross-section area. The results obtained showed rather constant values, suggesting that a coherent choice for $\xi_f/\left(\sqrt{g}A_0^{3/4}\right)$ should be around 10^{-10}. We also found that an efficient ratio ξ_f/ξ_p is around 10^{-6}.

3.3. Performance of the Models

The test case chosen for the comparison between CasADi and our algorithm is a rectangular channel (1 m wide) with a 100 m long sine wave bottom as shown in the following equation:

$$z(x) = 0.05\ \sin\left(\frac{\pi}{4}x\right) + 0.05 \tag{22}$$

A Manning friction formula with $n = 0.04$ s/m$^{1/3}$ was used to estimate friction losses. The downstream boundary condition is a water depth equal to 0.6 m. The unit discharge is uniform and is equal to 1 m^2/s. The precision parameters are as follows: $\xi_p = 10^{-6}$ m^2/s and $\xi_f = 10^{-8}$ m^2/s.

We compared the CPU time spent in the solving stages of our new algorithm and CasADi. The goal was to compare the evolution of the computation time with the number of nodes, rather than comparing absolute values. The results are given in Figure 13. They confirm that the computation time evolves almost linearly with the number of nodes when the new algorithm is used. This is not the case with CasADi: the computation time increases following a power law N^{α}, $\alpha > 1$. This is due to the matrix operations that CasADi has to perform. Increasing the number of nodes leads to a non-proportional increase in computation time. The speed up factor, which is the ratio between the CPU time spent when using our new algorithm and the CPU time spent with CasADi, ranges from in order of 10^0 to in order of 10^2 according to the number of nodes considered.

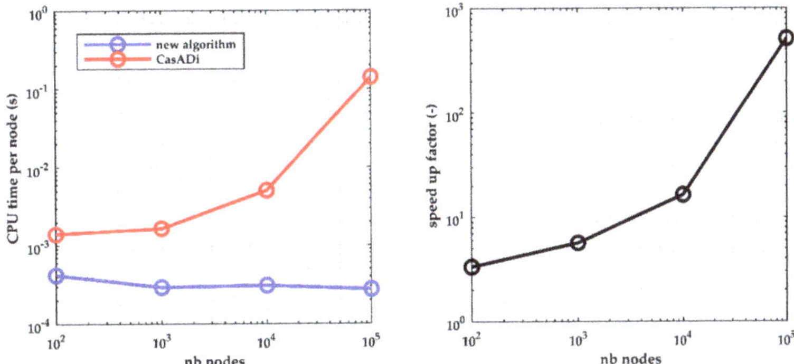

Figure 13. Comparison of the CPU time evolution with the number of nodes between our algorithm and CasADi and the associated speed up factor.

3.4. Real-World Application

The Romanche River in the French Alps is currently facing a number of significant changes. A new hydropower facility is being built in order to replace older power plants. In this context, the dam operator needs a fast computation routine in order to operate its facilities in an optimized and safe way. To do so, an unsteady 1-D model was implemented in Fortran and integrated in a Simulink S-Function in order to be compatible with the operator model [7]. Our new algorithm was used for the fast computation of an initial condition.

The studied part of the Romanche River stretches over 9.5 km (Figure 14), from the new Livet Dam to Gavet. The geometry and the calibration of the friction coefficients are detailed in [7]. The river was discretized with 191 nodes (50 m-long meshes). The downstream boundary condition was set at an elevation of 436.8 m. Two discharges were tested: 10 and 40 m^3/s. Since the details of the hydraulic results are of limited interest for this paper, we focused on the execution time and showed that the Froude number remains under 1 for both discharges (Figure 14). For the same machine as the one used earlier, the execution times (CPU time) are 0.018 s and 0.022 s, respectively ($\xi_f = 10^{-8}$ m^2/s and $\xi_p = 10^{-2}$ m^2/s). In comparison, the computation time of a five-minute full simulation was about 0.2 s. Given the length of the stretch and the celerity of a wave, the time for a wave to travel from downstream to upstream is of the order of 80 min. This means that approximately 3.2 s of computation time are required is order to simulate the propagation of a wave downstream to upstream. It is clear that the gain of time offered by our original technique is significant.

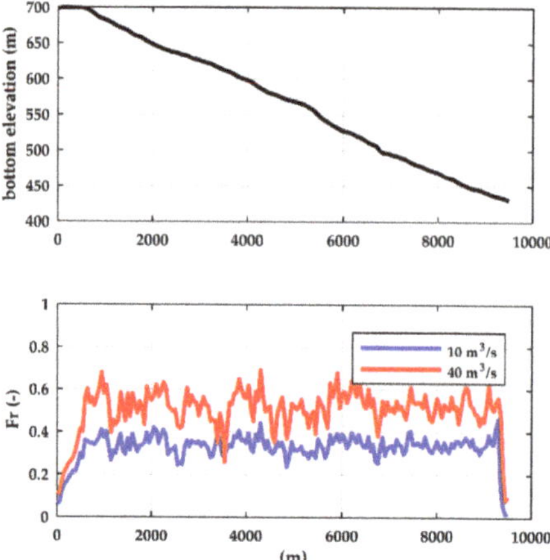

Figure 14. Bottom elevation and Froude number distribution for both simulated discharges.

4. Conclusions

Several innovations are introduced in this paper, including the use of the non-linear Krylov accelerator in open-channel flows, an evolutionary domain algorithm and the use of CasADi to solve steady 1-D flows. These improvements lead to an algorithm that is able to quickly solve steady open-channel flows. Therefore, optimization problems and uncertainty analyses that require many evaluations, become more tractable.

An original algorithm was implemented in order to significantly improve the computation time of a steady 1-D open-channel flow problem. It includes two main optimizing strategies: a non-linear Krylov accelerator and an evolutionary domain algorithm. This new algorithm was validated against the academic benchmarks of flows over a bump. The results showed a good agreement between the numerical and analytical values.

The performance of the suggested algorithm was evaluated against the non-linear optimization software CasADi. It showed good scalability properties. Indeed, the execution time of the proposed algorithm evolves linearly with the number of nodes. This is not the case with other techniques when the mesh is refined and/or when the number of nodes increase.

Finally, we demonstrated the capabilities of our algorithm in a real-world case. We used the optimized algorithm in order to compute quickly the initial condition required by the operational model for the Romanche River in France. Our technique was able to provide a steady state solution to the unsteady model in a very short period of time.

Author Contributions: Conceptualization, M.P. and P.A.; methodology, L.G. and P.A.; software, L.G. and P.A.; validation, L.G.; writing—original draft preparation, L.G.; writing—review and editing, P.A., M.P., B.D. and S.E.; supervision, M.P. and P.A. All authors have read and agreed to the published version of the manuscript.

Funding: This research received no external funding.

Conflicts of Interest: The authors declare no conflict of interest.

Computer Code and Software: The following software and codes were used for this paper: (a) WOLF was developed by the HECE research group (http://www.hece.ulg.ac.be/cms/) at the University of Liège since 2000 and is not freely available, (b) CasADi is freely available at https://web.casadi.org/, and (c) the routines used to test CasADi are freely available at https://gitlab.uliege.be/HECE/HydroCasADi.

References

1. Proust, S.; Berni, C.; Boudou, M.; Chiaverini, A.; Dupuis, V.; Faure, J.-B.; Paquier, A.; Lang, M.; Guillen-Ludena, S.; Lopez, D.; et al. Predicting the flow in the floodplains with evolving land occupations during extreme flood events (FlowRes ANR project). In Proceedings of the E3S Web of Conferences, Lyon, France, 17–21 October 2016; Volume 7.
2. Drab, A.; Riha, J. An approach to the implementation of European Directive 2007/60/EC on flood risk management in the Czech Republic. *Nat. Hazards Earth Syst. Sci.* **2010**, *10*, 1977–1987. [CrossRef]
3. Schwanenberg, D.; Becker, B.P.J.; Xu, M. The open real-time control (RTC)-Tools software framework for modeling RTC in water resources sytems. *J. Hydroinform.* **2014**, *17*, 130–148. [CrossRef]
4. Tayefi, V.; Lane, S.N.; Hardy, R.J.; Yu, D. A comparison of one- and two-dimensional approaches to modelling flood inundation over complex upland floodplains. *Hydrol. Process.* **2007**, *21*, 3190–3202. [CrossRef]
5. Horritt, M.S.; Bates, P.D. Evaluation of 1D and 2D numerical models for predicting river flood inundation. *J. Hydrol.* **2002**, *268*, 87–99. [CrossRef]
6. Cook, A.; Merwade, V. Effect of topographic data, geometric configuration and modeling approach on flood inundation mapping. *J. Hydrol.* **2009**, *377*, 131–142. [CrossRef]
7. Goffin, L.; Dewals, B.J.; Erpicum, S.; Pirotton, M.; Archambeau, P. Non-linear optimization of a 1-D shallow water model and integration into Simulink for operational use. In *Proceedings of the Sustainable Hydraulics in the Era of Global Change—4th European Congress of the International Association of Hydroenvironment Engineering and Research (IAHR 2016)*; Liege, Belgium, 27–29 July 2016, CRC Press/Balkema: Boca Raton, FL, USA, 2016; pp. 445–451.
8. Bourdarias, C.; Gerbi, S.; Gisclon, M. A kinetic formulation for a model coupling free surface and pressurised flows in closed pipes. *J. Comput. Appl. Math.* **2008**, *218*, 522–531. [CrossRef]
9. Paiva, R.C.D.; Collischonn, W.; Tucci, C.E.M. Large scale hydrologic and hydrodynamic modeling using limited data and a GIS based approach. *J. Hydrol.* **2011**, *406*, 170–181. [CrossRef]
10. Paz, A.R.; Bravo, J.M.; Allasia, D.; Collischonn, W.; Tucci, C.E.M. Large-scale hydrodynamic modeling of a complex river network and floodplains. *J. Hydrol. Eng.* **2010**, *15*, 152–165. [CrossRef]
11. Lai, X.; Jiang, J.; Liang, Q.; Huang, Q. Large-scale hydrodynamic modeling of the middle Yangtze River Basin with complex river-lake interactions. *J. Hydrol.* **2013**, *492*, 228–243. [CrossRef]
12. Remo, J.W.F.; Pinter, N. Retro-modeling the Middle Mississippi River. *J. Hydrol.* **2007**, *337*, 421–435. [CrossRef]
13. Biancamaria, S.; Bates, P.D.; Boone, A.; Mognard, N.M. Large-scale coupled hydrologic and hydraulic modelling of the Ob river in Siberia. *J. Hydrol.* **2009**, *379*, 136–150. [CrossRef]
14. Kerger, F.; Archambeau, P.; Erpicum, S.; Dewals, B.J.; Pirotton, M. A fast universal solver for 1D continuous and discontinuous steady flows in rivers and pipes. *Int. J. Numer. Methods Fluids* **2011**, *66*, 38–48. [CrossRef]
15. Sandric, I.; Ionita, C.; Chitu, Z.; Dardala, M.; Irimia, R.; Furtuna, F.T. Using CUDA to accelerate uncertainty propagation modelling for landslide susceptibility assessment. *Environ. Model. Softw.* **2019**, *115*, 176–186. [CrossRef]
16. Neal, J.C.; Fewtrell, T.J.; Bates, P.D.; Wright, N.G. A comparison of three parallelisation methods for 2D flood inundation models. *Environ. Model. Softw.* **2010**, *25*, 398–411. [CrossRef]
17. Lacasta, A.; Garcia-Navarro, P.; Burguete, J.; Murillo, J. Preprocess static subdomain decomposition in practical cases of 2D unsteady hydraulic simulation. *Comput. Fluids* **2013**, *80*, 225–232. [CrossRef]
18. Brodtkorb, A.R.; Sætra, M.L.; Altinakar, M. Efficient shallow water simulations on GPUs: Implementation, visualization, verification, and validation. *Comput. Fluids* **2012**, *55*, 1–12. [CrossRef]
19. Petaccia, G.; Leporati, F.; Torti, E. OpenMP and CUDA simulations of Sella Zerbino Dam break on unstructured grids. *Comput. Geosci.* **2016**, *20*, 1123–1132. [CrossRef]
20. Smith, L.S.; Liang, Q. Towards a generalised GPU/CPU shallow-flow modelling tool. *Comput. Fluids* **2013**, *88*, 334–343. [CrossRef]
21. Chaudhry, M.H. *Open-Channel Flow*; Springer Science & Business Media: Berlin/Heidelberg, Germany, 2007.
22. Bruwier, M.; Archambeau, P.; Erpicum, S.; Pirotton, M.; Dewals, B. Discretization of the divergence formulation of the bed slope term in the shallow-water equations and consequences in terms of energy balance. *Appl. Math. Model.* **2016**, *40*, 7532–7544. [CrossRef]

23. Franzini, F.; Soares-Frazão, S. Efficiency and accuracy of Lateralized HLL, HLLS and Augmented Roe's scheme with energy balance for river flows in irregular channels. *Appl. Math. Model.* **2016**, *40*, 7427–7446. [CrossRef]
24. Erpicum, S.; Dewals, B.; Archambeau, P.; Pirotton, M. Dam break flow computation based on an efficient flux vector splitting. *J. Comput. Appl. Math.* **2010**, *234*, 2143–2151. [CrossRef]
25. Broyden, C.G. A class of methods for solving nonlinear simultaneous equations. *Math. Comput.* **1965**, *19*, 577–593. [CrossRef]
26. Knoll, D.A.; Keyes, D.E. Jacobian-free Newton–Krylov methods: A survey of approaches and applications. *J. Comput. Phys.* **2004**, *193*, 357–397. [CrossRef]
27. Anderson, D.G. Iterative Procedures for Nonlinear Integral Equations. *J. ACM* **1965**, *12*, 547–560. [CrossRef]
28. Walker, H.F.; Ni, P. Anderson acceleration for fixed-point iterations. *SIAM J. Numer. Anal.* **2011**, *49*, 1715–1735. [CrossRef]
29. Saad, Y.; Schultz, M.H. GMRES: A generalized minimal residual algorithm for solving nonsymmetric linear systems. *SIAM J. Sci. Stat. Comput.* **1986**, *7*, 856–869. [CrossRef]
30. Carlson, N.N.; Miller, K. Design and application of a gradient-weighted moving finite element code I: In one dimension. *SIAM J. Sci. Comput.* **1998**, *19*, 728–765. [CrossRef]
31. Carlson, N.N.; Miller, K. Design and Application of a Gradient-Weighted Moving Finite Element Code II: In Two Dimensions. *SIAM J. Sci. Comput.* **1998**, *19*, 766–798. [CrossRef]
32. Calef, M.T.; Fichtl, E.D.; Warsa, J.S.; Berndt, M.; Carlson, N.N. Nonlinear Krylov acceleration applied to a discrete ordinates formulation of the k-eigenvalue problem. *J. Comput. Phys.* **2013**, *238*, 188–209. [CrossRef]
33. Wang, C.; Cheng, J.; Berndt, M.; Carlson, N.N.; Luo, H. Application of nonlinear Krylov acceleration to a reconstructed discontinuous Galerkin method for compressible flows. *Comput. Fluids* **2018**, *163*, 32–49. [CrossRef]
34. Andersson, J.A.E.; Gillis, J.; Horn, G.; Rawlings, J.B.; Diehl, M. CasADi—A software framework for nonlinear optimization and optimal control. *Math. Program. Comput.* **2018**, *11*, 1–36. [CrossRef]
35. Baayen, J.; Piovesan, T.; VanderWees, J. Optimization problems subject to the nonlinear semi-implicitly discretized Saint-Venant equations have a unique solution. *arXiv* **2018**, arXiv:1801.06507.
36. Aureli, F.; Maranzoni, A.; Mignosa, P.; Ziveri, C. A weighted surface-depth gradient method for the numerical integration of the 2D shallow water equations with topography. *Adv. Water Resour.* **2008**, *31*, 962–974. [CrossRef]

Article

Efficient Reservoir Modelling for Flood Regulation in the Ebro River (Spain)

Isabel Echeverribar [1,2,*], Pablo Vallés [1], Juan Mairal [1] and Pilar García-Navarro [1]

1 Fluid Mechanics, Universidad Zaragoza, I3A, Maria de Luna s/n, 50018 Zaragoza, Spain; 736453@unizar.es (P.V.); mairalascaso@unizar.es (J.M.); pigar@unizar.es (P.G.-N.)
2 Hydronia Europe, P° Castellana 95-15, 28046 Madrid, Spain
* Correspondence: echeverribar@unizar.es

Abstract: The vast majority of reservoirs, although built for irrigation and water supply purposes, are also used as regulation tools during floods in river basins. Thus, the selection of the most suitable model when facing the simulation of a flood wave in a combination of river reach and reservoir is not direct and frequently some analysis of the proper system of equations and the number of solved flow velocity components is needed. In this work, a stretch of the Ebro River (Spain), which is the biggest river in Spain, is simulated solving the Shallow Water Equations (SWE). The simulation model covers the area of river between the city of Zaragoza and the Mequinenza dam. The domain encompasses 721.92 km^2 with 221 km of river bed, of which the last 75 km belong to the Mequinenza reservoir. The results obtained from a one-dimensional (1D) model are validated comparing with those provided by a two-dimensional (2D) model based on the same numerical scheme and with measurements. The 1D modelling loses the detail of the floodplain, but nevertheless the computational consumption is much lower compared to the 2D model with a permissible loss of accuracy. Additionally, the particular nature of this reservoir might turn the 1D model into a more suitable option. An alternative technique is applied in order to model the reservoir globally by means of a volume balance (0D) model, coupled to the 1D model of the river (1D-0D model). The results obtained are similar to those provided by the full 1D model with an improvement on computational time. Finally, an automatic regulation is implemented by means of a Proportional-Integral-Derivative (PID) algorithm and tested in both the full 1D model and the 1D-0D model. The results show that the coupled model behaves correctly even when controlled by the automatic algorithm.

Keywords: reservoir model; numerical simulation; shallow water equations; PID regulation

Citation: Echeverribar, I.; Vallés, P.; Mairal, J.; García-Navarro, P. Efficient Reservoir Modelling for Flood Regulation in the Ebro River (Spain). *Water* **2021**, *13*, 3160. https://doi.org/10.3390/w13223160

Academic Editors: Anargiros I. Delis and Ioannis K. Nikolos

Received: 22 September 2021
Accepted: 12 October 2021
Published: 9 November 2021

1. Introduction

As extreme phenomena, flood events raise concern among governments, institutions and general society. The European Union has been developing plans and directives during the last decades focusing on the control of their impact [1]. River overflows cause the flooding of adjacent lands, urbanised areas and other infrastructures. Additionally, floods can also take human lives, as reported by the UN [2], specially in areas with poor prevention plans and a lack of predictive tools. Frequently, dams and reservoirs are present in river basins as hydraulic elements with different functions. Not only to ensure enough water supply for agricultural activities or energy production, but also as hydraulic structures for discharge adjustment and control during flood events. Basin authorities manage their operation focusing on available space in the reservoir, maximum acceptable downstream discharges, and peak arrival times.

In this context, the development of predictive tools that provide information about the temporal and spatial evolution of water level and discharge along a river during flood events can help to quantify the damage caused and has been widely addressed in last decades [3]. Some works are focused on urban areas coupling their overland models with sewer systems [4,5]. Some others are more oriented to large scale floods quantifying inundated areas in crops and surrounding fields [6,7], including components such as weirs, gates, dams and reservoirs [8,9]. Nowadays, there are even operational tools developed to simulate extremely large domains using massive parallel algorithms [10]. In either case, all models are then used to generate information and data for other models and tools that analyse and classify flood-prone areas [11,12]. Fast and efficient numerical models for the resolution of the equations that govern free surface flows in rivers have been developed and improved in recent years [6,7,10,13–17]. However, including a reservoir in the numerical model of the river reach is an additional difficulty. Even today, it is widely accepted that three dimensional models (3D) are computationally expensive and the phenomenon of a flood in a river at a large scale can be addressed by averaging the equations vertically (2D approximation) [6,7]. However, flood event simulations often involve large domains and long time scales, and practical applications require a compromise between spatial resolution and computational efficiency [18,19]. To achieve the necessary spatial resolution, in many cases quite fine computational grids are needed, so more data storage is required, proportionally increasing the number of operations and reducing the size of the time step allowed for explicit calculations. Therefore, flood risk evaluations are often performed considering the average in the cross section to reduce the phenomenon to a 1D approximation [13]. Finally, depending on terrain morphology, some particular river reaches might transport hydrographs almost immediately, as reservoirs.

On the other hand, reservoirs can be assessed with different approximations depending on the variables of interest. Reservoirs can be solved either as part of the river reach; this is, discretised. Alternatively, they can be considered as a storage volume with a constant level [20,21]. When the detailed phenomena that might occur within a reservoir are of interest, complex numerical models are developed. In [22], a 3D model based on the Reynolds-Averaged Navier-Stokes equations is used to study secondary currents and three dimensional behaviour of the velocity field. When the details of the 3D velocity field are not required but the longitudinal profile of the water surface is of interest, they can be incorporated as part of 1D discretised models [23]. Additionally, when some other additional phenomena must be simulated, such as eutrophization [24] or sediment transport [25], also a spatial discretisation of the reservoir is needed, whether in one or three dimensions. An alternative is an aggregated reservoir routing where only a volume balance is considered [26–28]. This may include runoffs, evaporation and some other mass exchanges. In any case, each reservoir must be specifically analysed. Aggregated models may be suitable, due to the representation of the reservoir as a unique volume, providing CPU times in the order of seconds [28], and a discretised model must compute as many operations as grid elements, leading to higher computational times [22]. However, the main disadvantage of these simplified approaches is that they do not represent in detail the flow behaviour of the river and floodplains [29–31].

Concerning optimization of the reservoir as a hydraulic structure, several works have focused on their hydropower potential. In [32], for instance, a linear optimization model with three different objective functions was implemented to automatically manage the reservoir in order to maximize total energy.

The aim of the present work is to couple recent research tools based on shallow water numerical models for flood forecasting with an aggregated model for the reservoir, developing a complete efficient simulation tool during flood events.

First, a comparison of the results obtained with a 2D and a 1D model in the middle reach of the Ebro river is carried out for validation purposes. Both, the 2D and the 1D model, are based on a finite volume scheme, which uses terrain data for the 2D mesh and 1D bathymetry creation. The 1D modelling is likely to lose the detail of the floodplain, but nevertheless the computational cost is expected much lower compared to the 2D model. Additionally, the particular nature of this reservoir, which is highly channelised, might turn the 1D discretisation into a more suitable option than the 2D discretisation. Therefore, the quality of the 1D results should be checked and the calculation times of both models should be compared.

Due to the particularity of the simulated section and the Mequinenza reservoir, which transports flood waves almost immediately, an aggregated alternative technique is applied. This approach formulates the reservoir flow globally by means of a volume balance model. Our focus is on checking if the results obtained are similar to those provided by the fully 1D model and comparing the computation times of both simulations. This later option is completed with a PID algorithm for regulation purposes.

2. Study Area

The Ebro River basin is one of the largest drainage areas in the Iberian Peninsula, as seen in Figures 1a,b, where the Ebro River represents the biggest river in Spain. In this work, a stretch of this Ebro River is simulated solving the Shallow Water Equations (SWE). The simulation model, delimited in dashed line in Figure 1c, covers the area between the city of Zaragoza and the Mequinenza dam, encompassing 721.92 km^2. Between the inlet and outlet locations, there are 221 km of river bed of which the last 75 km belong to the Mequinenza reservoir, where the dynamics of the river changes to be nearly at rest. During the entire stretch, the river descends from 208 m.a.s.l. of the elevation in Zaragoza up to approximately 60 m.a.s.l. at the bottom of the riverbed in the Mequinenza dam, leaving an average slope of 6 per 10,000.

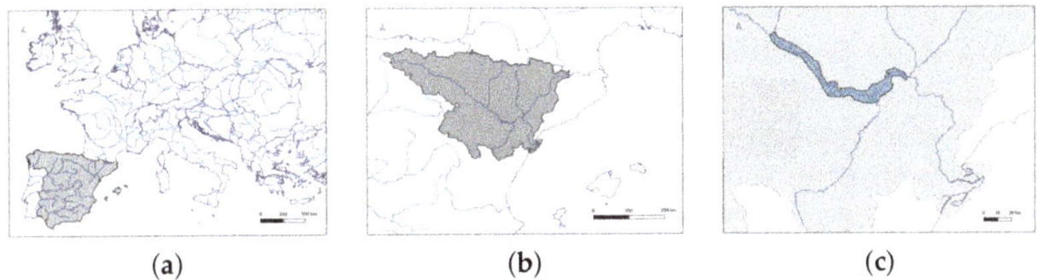

Figure 1. Location of Spain in Europe (**a**); location of the Ebro River basin in Spain (**b**) and location of the computational domain of the study in the basin (**c**).

The Ebro river is managed by the Ebro River Authority (CHE, www.chebro.es), which controls, rules and prepares the reports of the basin (http://www.chebro.es/contenido.visualizar.do?idContenido=14093&idMenu=3048; accessed on 20 October 2021). CHE monitors the evolution of the flow discharge and water levels at a few control stations along the river course storing data every 15 min. Figure 2 represents the Ebro River reach simulated in this work. In the figure, the most important cities and the CHE gauging stations available for data comparison are marked. The represented domain coincides with the 2D domain used for simulations. The gauging stations in this reach are located in Pina, Villafranca and Gelsa. Additionally, a point of estimation exist near the Mequinenza dam. Each of these stations has an official label that can be seen in the same figure. This region is of special interest due to its agricultural activity, and frequently suffers flooding with important damages. It is a river reach where two different parts can be identified: the first part of the region is characterised by marked meanders and large flooding areas; the second part, around 75 km of the reach, is dominated by the large Mequinenza reservoir provided with vertical walls. The dynamics of the river changes in the reservoir: its velocity is reduced until water is practically at rest and flood waves are transported almost instantly.

The Mequinenza reservoir, the largest in the entire region, covers a surface area of about 7540 hectares, with a maximum capacity of 1530 hm^3 at a maximum normal surface level of 121 m.a.s.l. The reservoir is exploited for hydroelectric production and irrigation to nearby agricultural areas. At the same time, with its 124 m crest above sea level and its 6 gates, the dam is used to regulate the water storage in order to dump peak discharge during floods and to guarantee hydroelectric generation. At the reservoir, there is only a measurement point for water level that is transformed by CHE into a discharge value through volume estimations. CHE uses two different approaches for discharge estimation, so that the results in the Mequinenza reservoir are compared not with observed but with 2 different estimated data.

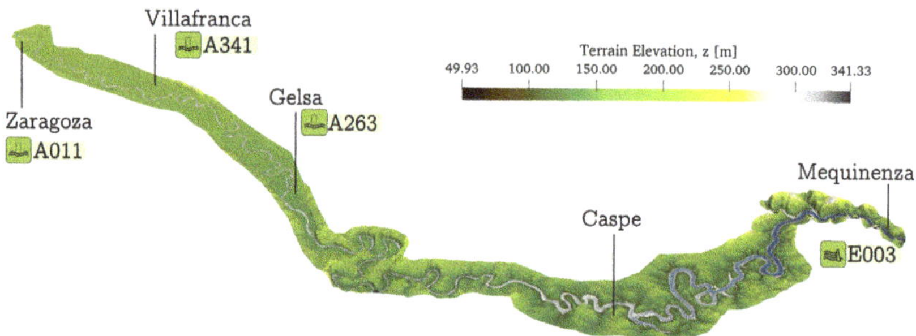

Figure 2. Representation of the 2D simulation domain of the Ebro River with the most important cities and gauging stations of CHE. The labels correspond to the official names of the gauging stations.

Two historical events of the Ebro River, the 2015 and the 2018 floods, have been identified as relevant. Information concerning discharge hydrographs as well as time evolution of the water surface level are available at the gauging stations. Additionally, the European Emergency System (EMS) provides data of flooded area extensions, as seen in Figure 3 (https://emergency.copernicus.eu/; accessed on 20 October 2021). This helped to choose the domain extension setting the boundaries far enough not to interfere the flow.

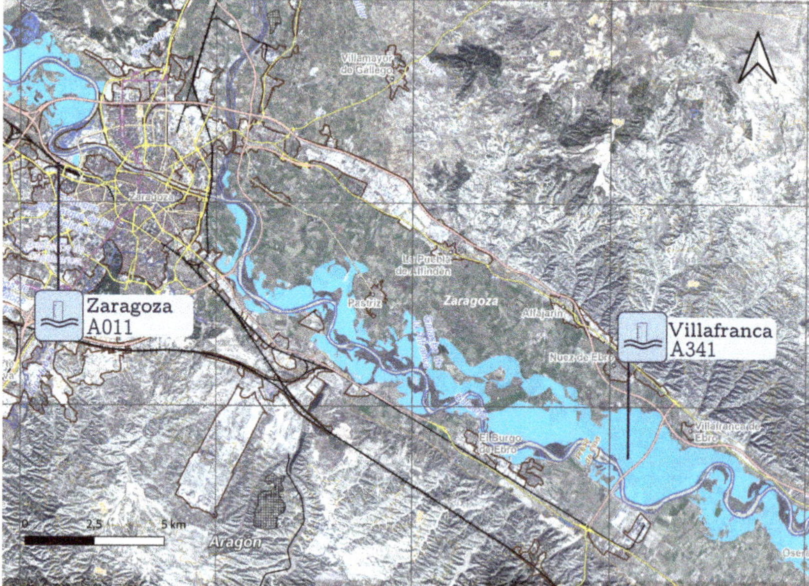

Figure 3. Zoom view of an ortophoto with measured extension of the flooded area (blue) in 2018 flood event.

3. Methodology

Derived from the Navier-Stokes equations by depth averaging and assuming hydrostatic pressure, the Shallow Water Equations (SWE) can be considered to govern the free surface flow of a river.

3.1. Two Dimensional (2D) Model

The 2D model can be compactly formulated as

$$\frac{\partial \mathbf{U}}{\partial t} + \frac{\partial \mathbf{F}(\mathbf{U})}{\partial x} + \frac{\partial \mathbf{G}(\mathbf{U})}{\partial y} = \mathbf{S}(\mathbf{U}) \tag{1}$$

with:

$$\mathbf{U} = \begin{pmatrix} h \\ hu \\ hv \end{pmatrix} \quad \mathbf{F}(\mathbf{U}) = \begin{pmatrix} hu \\ hu^2 + gh^2/2 \\ huv \end{pmatrix} \tag{2}$$

$$\mathbf{G}(\mathbf{U}) = \begin{pmatrix} hv \\ huv \\ hv^2 + gh^2/2 \end{pmatrix} \quad \mathbf{S}(\mathbf{U}) = \begin{pmatrix} 0 \\ gh(S_{ox} - S_{fx}) \\ gh(S_{oy} - S_{fy}) \end{pmatrix} \tag{3}$$

in terms of the water depth, h, the depth averaged unit discharges hu and hv in the x and y directions respectively. The slopes S_{ox} and S_{oy} are the two components of the bottom surface gradient $z_b(x,y)$:

$$S_{ox} = -\frac{\partial z_b}{\partial x} \quad S_{oy} = -\frac{\partial z_b}{\partial y}; \tag{4}$$

and S_{fx} and S_{fy} represent friction slopes, that are here formulated as:

$$S_{fx} = \frac{n^2 u\sqrt{u^2+v^2}}{h^{4/3}} \quad S_{fy} = \frac{n^2 v\sqrt{u^2+v^2}}{h^{4/3}} \tag{5}$$

where n stands for the semiempirical Manning friction coefficient ([33]).

3.2. One Dimensional (1D) Model

When the equations are averaged over the cross sectional area of the flow, a 1D model is obtained, representing changes along the longitudinal direction of the river. The obtained system is analogous to the 2D system, with a mass conservation equation and a linear momentum equation along the river channel:

$$\frac{\partial \mathbf{U}}{\partial t} + \frac{\partial \mathbf{F}(\mathbf{U})}{\partial x} = \mathbf{S}(\mathbf{U}) \tag{6}$$

with:

$$\mathbf{U} = \begin{pmatrix} A \\ Q \end{pmatrix} \quad \mathbf{F} = \begin{pmatrix} Q \\ Q^2/A + gI_1 \end{pmatrix} \quad \mathbf{S} = \begin{pmatrix} 0 \\ g[I_2 + A(S_0 - S_f)] \end{pmatrix} \tag{7}$$

where Q stands for transversal discharge, A is the cross section wetted area and I_1, I_2 are hydrostatic pressure integrals. S_0 is the bottom slope along the longitudinal coordinate of the channel:

$$S_0 = -\frac{\partial z_b}{\partial x} \tag{8}$$

and S_f is the friction slope, that is also formulated through the Manning law as:

$$S_f = \frac{Q^2 n^2}{A^2 R^{4/3}} \tag{9}$$

where R is the hydraulic radius, defined as $R = A/P$, being A the wetted area and P the wetted perimeter. Finally, the Manning friction coefficient is obtained empirically ([33]).

3.3. Finite Volume Model for the 1D Flow Equations

In this work, an explicit upwind first order finite volume method is used for both systems of equations ([34–36]). The systems of Equations (1) and (6), can be generally expressed as:

$$\frac{\partial \mathbf{U}}{\partial t} + \vec{\nabla}\mathbf{E} = \mathbf{S} \tag{10}$$

where $\mathbf{E} = \begin{pmatrix} \mathbf{F} \\ \mathbf{G} \end{pmatrix}$ in the 2D model, **F** and **G** are given by (2) and (3). Moreover, $\mathbf{E} = \mathbf{F}$ in the 1D model, with **F** given by (7). When integrating (10) into a control volume or cell, Ω and applying the divergence theorem, the following expression is obtained:

$$\frac{d}{dt}\int_{\Omega} \mathbf{U}\, d\Omega + \oint_{\partial\Omega} \mathbf{E}(\mathbf{U}) \cdot \hat{n}\, dl = \int_{\Omega} \mathbf{S}(\mathbf{U})\, d\Omega \tag{11}$$

where $\hat{n}$ is the outward unit vector in the normal direction to the volume Ω. From this, the 1D approach is next developed and details for the 2D approach can be found in [7,37]

Due to the hyperbolic character of the 1D equations, the numerical scheme used to solve them is based on the Jacobian matrix of the fluxes:

$$\mathbf{J} = \frac{\partial(\mathbf{E}\cdot\hat{n})}{\partial \mathbf{U}} \xrightarrow{1\text{D}} \mathbf{J} = \frac{\partial \mathbf{F}}{\partial \mathbf{U}} = \begin{pmatrix} 0 & 1 \\ c^2 - u^2 & 2u \end{pmatrix} \tag{12}$$

whose eigenvalues are:

$$\lambda_1 = u - c \quad \lambda_2 = u + c \tag{13}$$

with u and c given by:

$$u = \frac{Q}{A} \quad c = \sqrt{g\, A/B} \tag{14}$$

being A the cross section and B the free surface width. The celerity c characterizes the speed of the infinitesimal surface deformation waves defining the dimensionless Froude number $Fr = \frac{u}{c}$.

Following [16,35,38], the final updating scheme for a cell i of the domain in the time t^{n+1} takes into account the contributions of neighbour cells containing fluxes and source terms as:

$$\mathbf{U}_i^{n+1} = \mathbf{U}_i^n - \frac{\Delta t_{1D}}{\Delta x}\left[\sum_{m=1}^{2} (\tilde{\lambda}^{+}\tilde{\gamma}\tilde{\mathbf{e}})_{i-1/2}^{m} + \sum_{m=1}^{2} (\tilde{\lambda}^{-}\tilde{\gamma}\tilde{\mathbf{e}})_{i+1/2}^{m}\right]^n \tag{15}$$

being $\tilde{\lambda}$ and $\tilde{\mathbf{e}}$, respectively, the eigenvalues and eigenvectors of the Jacobian matrix of the flux, $\tilde{\mathbf{J}}$, linearised on the cell edge. Additionally, Δx stands for the cell size. The upwind scheme sends the information to the wave propagation direction through the eigenvalues and their sign:

$$\tilde{\lambda}_{i+1/2}^{\pm m} = \frac{1}{2}(\tilde{\lambda} \pm |\tilde{\lambda}|)_{i+1/2}^{m} \tag{16}$$

This scheme for the ordinary computational cells must be complemented with proper initial and boundary conditions. The numerical scheme is stabilised by dynamically limiting the time step size, Δt_{1D}, with the CFL condition:

$$\Delta t_{1D} = \text{CFL}\, \min_{m,k}\left(\frac{\Delta x}{|\tilde{\lambda}_k^m|}\right) \tag{17}$$

where $0 < \text{CFL} \leq 1$ [39].

3.4. Reservoir Model

In the context of the 1D shallow water model, the reservoir can be modelled with two different approaches. In both cases, sketched in Figure 4, the upstream river reach is discretised with a 1D finite volume method. However, the two approaches differ in reservoir representation:

(a) 1D model: Fully discretised as the rest of the domain and solving the flow at each cell (see Figure 4a).
(b) 1D-0D model: assuming a lake-at-rest condition within the reservoir and embedding it into an aggregated model (0D) (see Figure 4b).

As depicted in Figure 4, the aim of approach (b) is to remove the computational cells needed for the reservoir. Using the sketch as an example, while the fully 1D distributed model encompasses from $x_L = 0$ to $x_L = L$, the coupled 1D-0D model has a discretised domain only from $x_L = 0$ to $x_L = L'$, so that the computational cost is reduced.

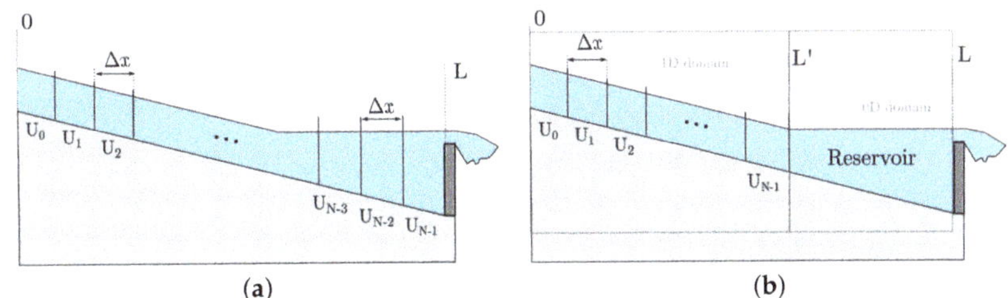

Figure 4. Representation of the two different approaches for reservoir representation: 1D distributed discretisation as the rest of the domain (**a**); or coupling the 1D model of the river with an aggregated 0D model of the reservoir (**b**).

In near-rest flows, such as those in a reservoir, the velocity field is negligible so that it is likely that the flow behaviour is properly solved only with volume balance. This represents a 0D approximation. The *Modified Puls Method* [21] is based on the hypotheses that the flow surface is always horizontal, the stored volume in the reservoir can be formulated as a function of water level ($V = V(H)$) and the outlet discharge can also be expressed as a water level function ($Q_{out} = Q_{out}(H)$). Thus, the volume variation is given by the difference between the reservoir inlet, Q_{in}, and outlet, Q_{out}, as:

$$\frac{dV}{dt} = Q_{in} - Q_{out} \tag{18}$$

Discretising Equation (18) in time by assuming $\Delta V = S(H^n)(H^{n+1} - H^n)$ with S the free surface reservoir area ($S = f(H)$) leads to:

$$H^{n+1} = H^n + \frac{\Delta t}{S(H^n)}\left(\frac{Q_{in}^{n+1} + Q_{in}^n}{2} - \frac{Q_{out}^{n+1} + Q_{out}^n}{2}\right) \tag{19}$$

When combining this formulation with the 1D model to lead to the 1D-0D model, the water level calculated with expression (19) is set at L' (Figure 4b). It is important to note, that the resolution of (18) requires knowing $Q_{in}(t)$ and the relation between Q_{out} and volume V at the reservoir. The inflow discharge to the reservoir is directly given by the computation at the last cell of the 1D model. On the other hand, the reservoir outflow discharge depends on the geometry and characteristics of the dam. In the present work the downstream boundary condition, either for pure 1D or for 1D-0D model, is based on a weir/dam law of the form [40]:

$$Q^{n+1} = \frac{2}{3}\sqrt{2g}bCH_w^{4/3} + \frac{8}{15}\sqrt{2g}tan\left(\frac{\theta}{2}\right)CH_w^{5/2} \tag{20}$$

where $H_w = H - h_{Crest}$ is the water depth above the weir crest. Assuming a trapezoid shape, b is the width of the minor length of the horizontal sides of the weir and θ the opening angle of the trapezoid. Finally, C is an energy loss coefficient here taken as $C = 0.611$ [9,26].

It is important to note that for both approaches, the full 1D model and the 1D-0D model, the 1D river reach upstream the reservoir must be identically discretised, this is, using the same Δx, so that the analysis can show the differences provoked by the reservoir model.

3.5. PID Regulation

The Mequinenza dam gates can be manually operated at present according to energy, agricultural or safety criteria. In this work, a control Proportional-Integral-Differential (PID) algorithm is implemented to show the possibility to dynamically include in the simulation model the control of the gate opening during a flood. In particular, a specific maximum reservoir surface level is set as target in the automatic algorithm so that the gate opening must change under discharge variations during the flood event.

The PID controller computes the error between the predicted value of the variable water surface level and the stated reference value, and uses it to compute a change on the free parameter, gate opening, using an algorithm based on:

- Proportional term: Expresses a proportionality between the required action and the error.
- Integral term: The required action takes into account the time integral of the error over a given period.
- Derivative term: The controller actuation is formulated from the time derivative of the error.

The equation that describes those PID terms is:

$$h_{Crest}(t) = K\left\{e(t) + \frac{1}{T_i}\int_0^{T_i} e(t)\,dt + T_d\,\frac{de(t)}{dt}\right\} = P + I + D \tag{21}$$

where $e(t)$ is the control error ($e(t) = H_{ref} - H(t)$), H_{ref} is the reference value (or *setpoint*) of water level and $H(t)$ is the current water level at time t. K is the proportional coefficient, T_i and T_d are integration and derivative times, respectively.

Equation (21) is discretised as:

$$h_{Crest}(t^n) = \alpha_1\,K\left(1 + \frac{T_s}{T_i} + \frac{T_d}{T_s}\right)e(t^n) - \alpha_2\,K\left(1 + \frac{2T_d}{T_s}\right)e(t^{n-1}) + \alpha_3\,K\,\frac{T_d}{T_s}\,e(t^{n-2}) \tag{22}$$

where $e(t^n) = H_{ref}(t^n) - H(t^n), e(t^{n-1}) = H_{ref}(t^{n-1}) - H(t^{n-1})$ and $e(t^{n-2}) = H_{ref}(t^{n-2}) - H(t^{n-2})$, being n the current time step. Parameters α_1, α_2 and α_3 are the weights given to each of the time steps that are included on the controller operation. Parameter T_s stands for the sampling period for the input data to the algorithm.

The values for K, T_i, T_d and T_s directly affect the h_{Crest} evolution and, thus, have an effect on the speed at which the controlled variable (i.e. water level) reaches the *setpoint*. Therefore, proper determination of those parameters is essential to optimize and stabilize the algorithm. Otherwise, the controller could lead to extreme gate opening values hence destabilizing the system.

4. Model Application

4.1. Discretisation of the Domain

The Ebro River reach has been first simulated to validate the 1D model comparing with results obtained with the 2D model [7] applied to the same stretch. In both cases the discretisation of the reservoir is included within the computational grid. The aim is to evaluate if the 1D approach provides reliable results improving the efficiency of the 2D model. Once the 1D model is demonstrated to be reliable enough, the coupled 1D-0D model and the control algorithm are evaluated.

The domain discretisation is different depending on the numerical scheme. In the 2D model, the mesh is an unstructured triangulation of the (x, y) domain with piecewise uniform values of terrain elevation and roughness. The triangles can be of variable size and adapt to the terrain topography. The mesh used was generated from a DTM in RASTER format and included 949,445 triangular elements. In the 1D model, the domain is discretised into a set of cells separated by cross sections along the riverbed evenly spaced at a distance Δx.

The 1D mesh was generated from topographic information of the field compound by 433 cross sections. Among them, 100 sections are within the reservoir region (as in example in Figure 5). The lateral span of the sections must capture the shape of the river bed to avoid losing information relevant to the evolution of the variables but avoiding overlapping. It is of vital importance that the sections are always normal to the river in curved regions, as seen in Figure 5. Finally, a 1D mesh of 2000 cells is obtained. Additionally, a uniform roughness coefficient n = 0.032 is chosen ([33]). A steady flow with a discharge value that matches the value at the initial time of the inlet hydrograph is set as initial condition. The upstream boundary condition is a hydrograph, while the downstream boundary condition is a spillway condition, which represents the presence of the Mequinenza dam.

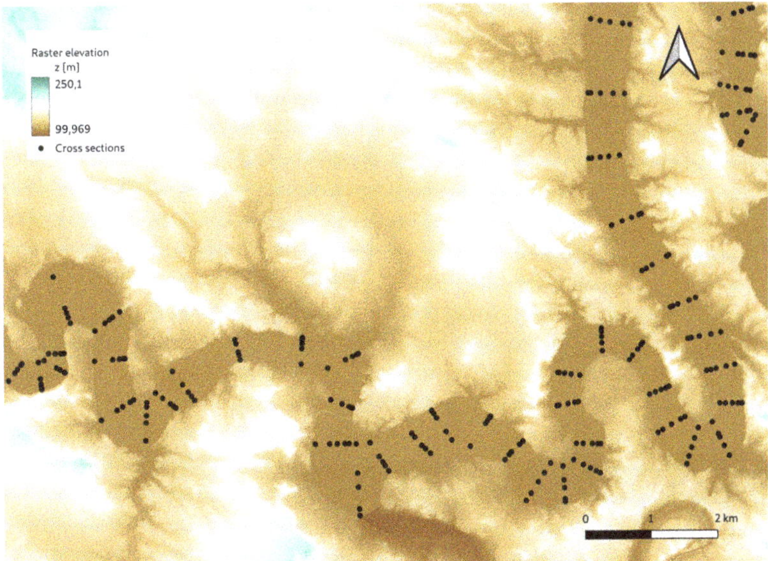

Figure 5. Top view of several sections over the raster with different river meanders.

4.2. Performance Analysis of the 1D and 2D Models

Two historical events of the Ebro River, the 2015 and the 2018 floods, have been simulated with both the 1D and the 2D models. The 2015 inlet hydrograph, obtained from the Zaragoza gauging station (see Figure 2), is set in Zaragoza as inlet boundary condition and can be seen in Figure 6. The comparison between results obtained with both simulation models and real observation data for this event can be seen in Figures 7 and 8. They show, respectively, the time evolution of discharge and water level at Gelsa (A263) station and Mequinenza dam (E003). It is important to note that the data provided by the CHE gauging station are for water depth (h) and not for water level ($H = h + z_b$), and the actual bed elevation of the station is unknown.

It can be seen in Figures 7 and 8 that the 1D and 2D simulations produce remarkably similar data which follow the tendency of the actual data. However, neither of the two models is able to reproduce the detail of the curves at $t = 380$ h. This is possibly due to a dynamic change in the terrain such as a levee breach not included in the static terrain representation of the models owing to the lack of available information. It should be noted that, in the first 250 h, it is the 1D model which offers data closer to the reality. This suggests that the flow was more channeled in this period of time and the overflow is estimated by the 2D model too early. From $t = 300$ h it is the 2D model which provides a behaviour closer to the real one, possibly due to the fact that, near the peak flow, the floodplains are inundated.

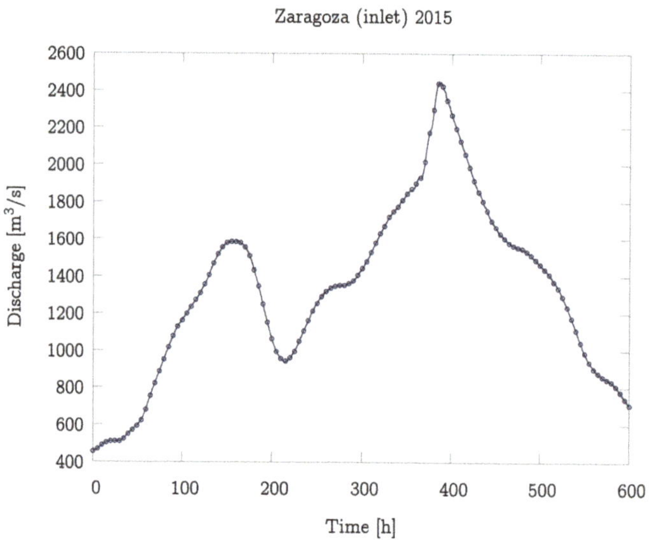

Figure 6. Inlet hydrograph for the Ebro River flood event in 2015 in Zaragoza (A011).

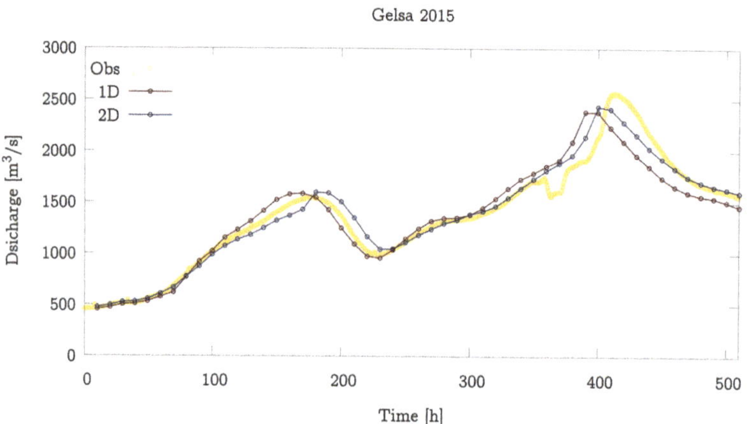

Figure 7. *Cont.*

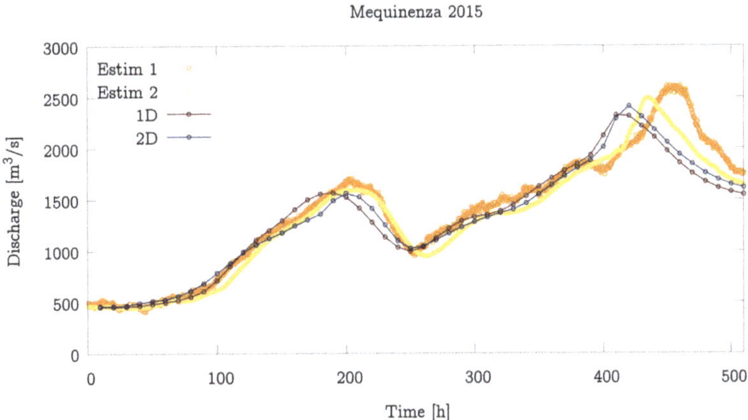

Figure 7. Discharge temporal evolution comparison between 1D model, 2D model and observation at Gelsa (A263) gauging station (**upper**) and comparison between models and estimations at Mequinenza dam (E003) (**lower**) for 2015 event.

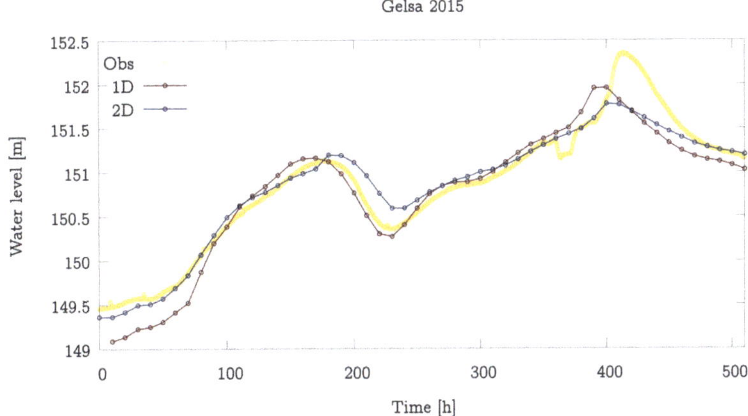

Figure 8. Water level temporal evolution comparison between 1D model, 2D model and observation data at Gelsa (A263) gauging station for the 2015 event.

Figure 9 shows the 2018 inlet hydrograph used as upstream boundary condition. Figure 10 displays the discharge time evolution as computed with the 1D and the 2D models together with the observation at Gelsa (A263) gauging station (upper) and Mequinenza dam (E003) (lower) for this event. The time evolution of the surface water level at Villafranca gauging station (the only measured variable) is displayed in Figure 11. The evolution of the water levels predicted by the two models is again quite similar to that of the real data until approximately $t = 100$ h. Around this time, the measured data suffer a slight decrease not predicted by the 2D model.

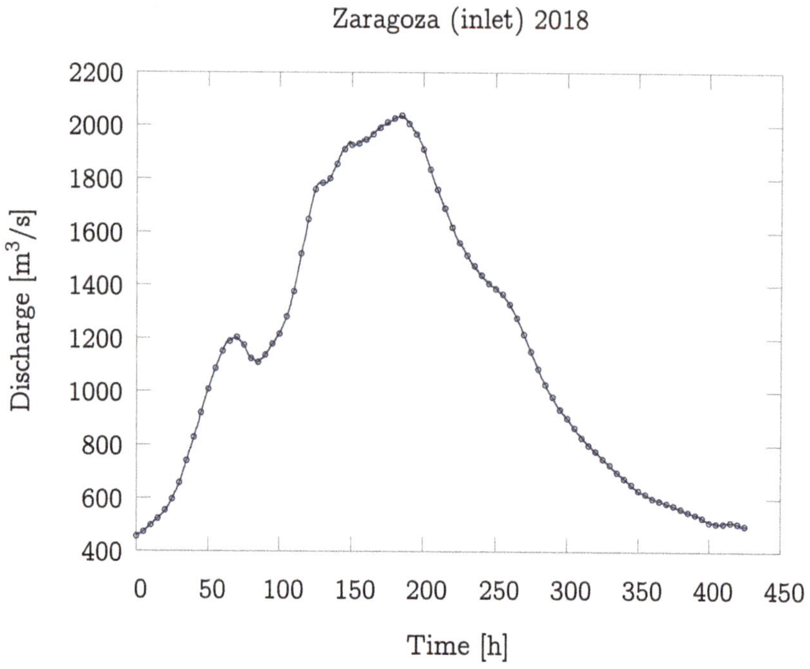

Figure 9. Inlet hydrographs for the Ebro River flood event in 2018 in Zaragoza (A011).

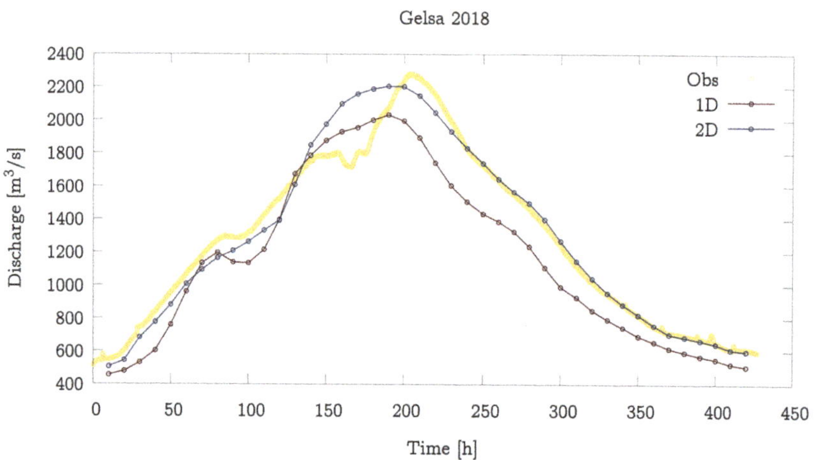

Figure 10. *Cont.*

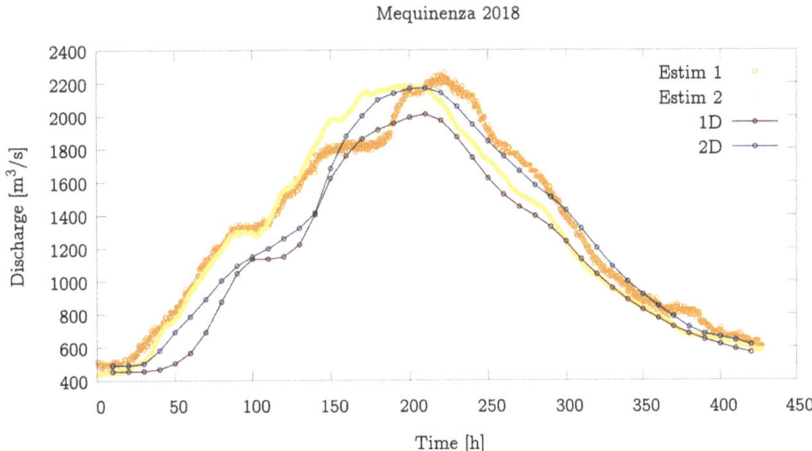

Figure 10. Discharge temporal evolution comparison between the 1D model, the 2D model and the observation at Gelsa (A263) gauging station (**upper**) and comparison between models and estimations in Mequinenza dam (E003) (**lower**) for the 2018 event.

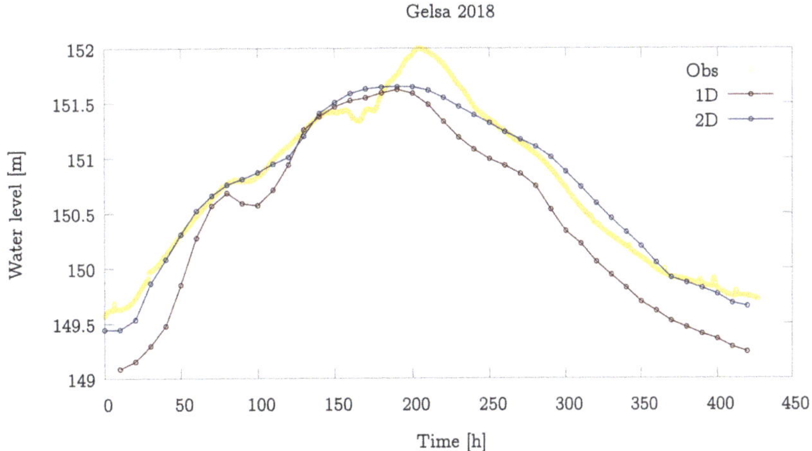

Figure 11. Water level temporal evolution comparison between the 1D model, the 2D model and the observation in Gelsa (A263) gauging station for the 2018 event.

Regarding computational times, for the 2015 case the 1D model required 511 s to compute the full event, while the 2D model took 47 h, as seen in Table 1. These computations were performed with High Performance Computing (HPC) techniques for the 2D model. In particular, a NVIDIA GeForce GTX 1080 Ti GPU was used to compute the 2D cases, while a simple paralellization into 8 Intel Xeon X5650 CPU's was necessary for the 1D computations. For the 2018 event, which is a bit shorter, the computational time required by the 1D model was 364 s, while that of the 2D model was 23.8 h.

Table 1. Comparison of computational times for the two flood events simulated in the Ebro River.

Event	Duration	2D GPU Time	1D CPU Time
2015	600 h	47 h	511 s
2018	430 h	23.8 h	364 s

4.3. Performance Analysis of the 1D and 1D-0D Models

The comparison between the full 1D model and the 1D-0D model is next carried out for the 2015 flood event. The discretisation of the full 1D model is the same used in the former subsection, while the 1D-0D model is characterised by a partial discretisation of the domain embedding the reservoir zone within the outlet boundary condition. In this last case, only the first 352 sections and 1511 mesh cells are necessary for the discretised part. This is, following Figure 4, for the river reach from $x_L = 0$ to $x_L = L'$.

Figure 12 represents the river profile for different times of both the full 1D model (fine and dark lines) and the 1D-0D model (wider and light lines). Bottom elevation, z_b, as well as water level, $h + z_b$, profiles are represented for both models. The initial condition can be seen at the upper picture of the figure, with a low water depth profile upstream the reservoir area, where the level remains uniform and the water depth increases. The middle picture corresponds to $t = 300$ h, when the discharge is increasing and a higher water depth can be seen. The lower picture coincides with the discharge peak of the inlet hydrograph (see Figure 6), reaching the highest value of water level. In the three cases, the level reached by the last cell of the 1D-0D model is almost the same as the value of the full 1D model, which discretises the whole reservoir. Therefore, it can be said that embedding the reservoir in the outlet boundary condition provides very similar results to those of a full model, without the necessity of such amount of computational cells.

As illustrated in Figure 12, the section located at $x_L = L'$ is not exactly the beginning of the reservoir, as the length of the reservoir varies throughout the simulation depending on the level of the water surface. For that reason, this section is chosen displaced forward ensuring that it always belongs to the reservoir. Thus, for high level values, there is part of the reservoir being discretised and also simulated by the 1D numerical scheme.

The temporal evolution of the water level at $x_L = L'$ can be seen in Figure 13. Although there is a very good agreement, the figure shows that the level of the 1D-0D model is slightly below the value of the full model at that location ($x_L = L'$). This is because there is not a uniform level in the entire reservoir and the 1D model represents this behaviour. However, the 0D model assumes a constant level in the reservoir that matches quite exactly the level at the end of the reservoir in the 1D model ($x_L = L$). In addition, the lag previously obtained by the model 1D-0D is no longer present. The reservoir surface area function, $S(H)$, corresponds to the entire reservoir. However, as part of the reservoir is being discretised by the 1D model, the boundary condition of the 1D-0D model causes a slower evolution of the level, as it is considering that there is a larger modeled reservoir than there should be and, therefore, it overestimates the value of $S(H)$.

The computational times of these simulations can be seen in Table 2. It can be seen how the 1D-0D model allows for considerable time reduction due to the due to the absence of the cells representing the reservoir. These results show that the coupled model results accurate enough to predict water levels along the river providing a performance improvement.

Table 2. Computational times for 2015 flood event simulated in the Ebro River with the 1D-0D model and the pure 1D model.

Event	Duration	Pure 1D	1D-0D
2015	600 h	511 s	196 s

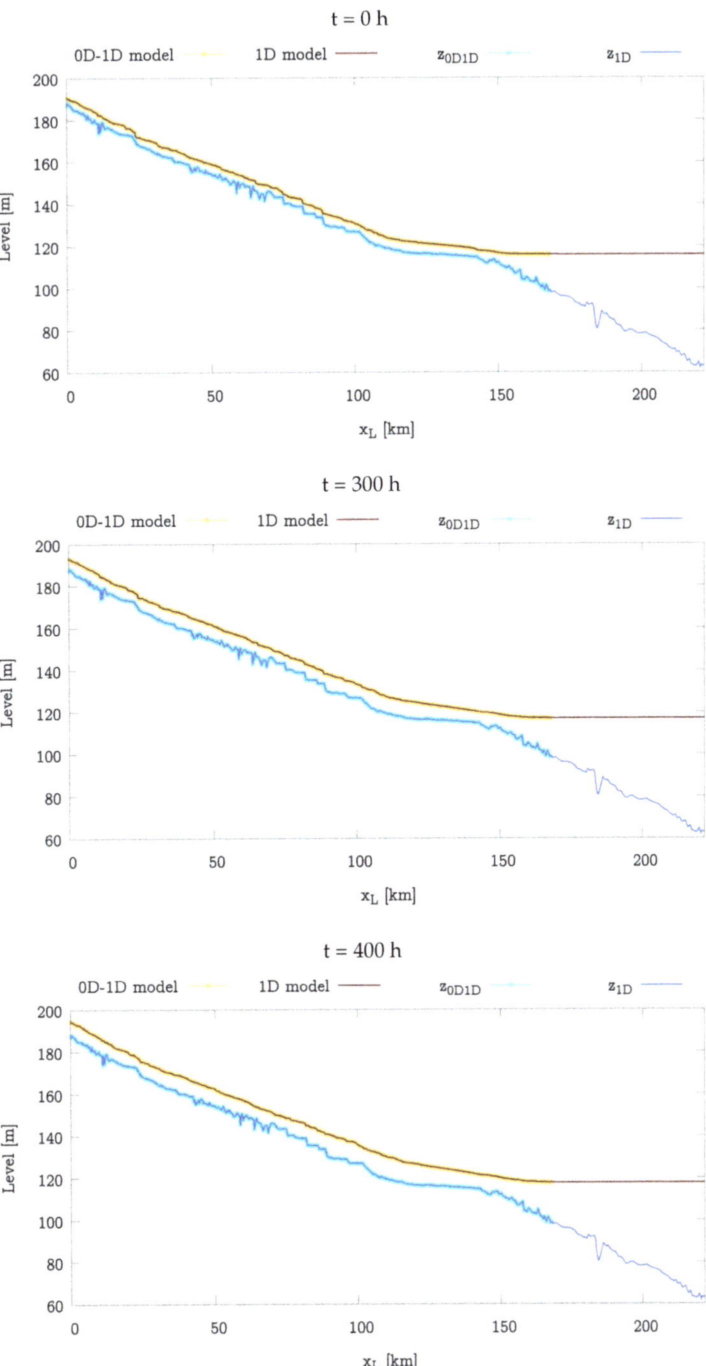

Figure 12. Longitudinal profile of bottom level, z, and water surface elevation (WSE) at different times computed with the 1D model and the coupled 1D-0D model for the 2015 case.

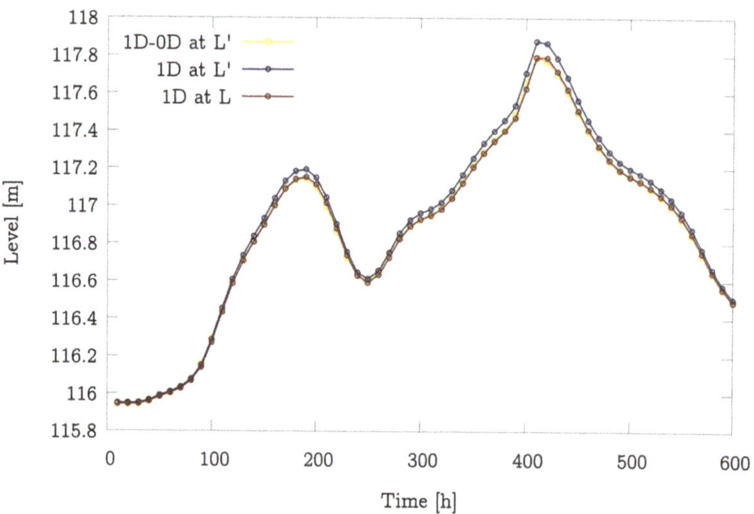

Figure 13. Comparison of computed water surface elevation (WSE) at x_L=L′ using 1D model and 1D-0D model and computed WSE at x_L=L using 1D model.

4.4. Performance Analysis of the 1D and 1D-0D Models Including DAM Regulation

Once the coupled 1D-0D model has been proved to perform properly, a dam regulation algorithm is implemented in both the coupled and the full 1D model. This is, both models include in their outlet boundary condition a PID algorithm that updates the dam crest, h_{crest}, to approach or to maintain a reference water level or *setpoint*, regardless of the inlet discharge.

For that purpose, both models have been discretised exactly as in the preceding section. The parameters used in the PID controller must be first calculated and validated. In this work, the chosen values are $K = 11576$, $T_i = 12$ s, $T_d = 3$ s, $T_s = 1000$ s, $\alpha_1 = 0.5$, $\alpha_2 = 0.3$ and $\alpha_3 = 0.2$. Those values were obtained following [41]. The dam movement is limited by $v_{max} = 0.01$ m per time interval and the water surface level is limited by a maximum and minimum value of 115 m and 105 m respectively. The target water surface level is $H_{ref} = 112$ m. The comparison is done at $x_L = L'$ (see Figure 4).

It is worth mentioning that variations of dam crest, h_{crest}, are a practical representation of variations in cross section area of spillways. In reality, dams can not change their crest, but the gate opening of their structure. However, the discharge law implemented would be the same and the dam crest results in a very representative parameter of the dam opening.

The Figure 14 shows the temporal variation of water level at $x_L = L'$ for both models. Besides that, the figure also represents the time evolution of the dam crest throughout the simulation. At the same time, Figure 15 depicts, also for both models, the temporal evolution of outlet discharge. It worth mentioning that this discharge is the flow rate passing through the dam.

Figure 14 shows that, at the beginning of the simulation, the level of the reservoir is much lower than the *setpoint* ($H_{ref} = 112$ m), so the dam crest is at its maximum height preventing the water leaking (see Figure 15) and provoking an increase of water level. Once the reference is reached, the dam crest varies to maintain a constant water level while the inlet flow rate changes. At that time, the outflow hydrograph tends to resemble the inflow rate.

The time evolution of h_{Crest} displayed in Figure 14 is rather similar for both models, reaching the target value in a short time. It is worth noting that the 1D model reaches the objective earlier than the 1D-0D model. This is provoked by the lower level obtained with the 1D-0D model at x_L=L′ (see Figure 4) in comparison with that computed with the 1D model. This causes a delay in the reservoir filling up to the *setpoint*.

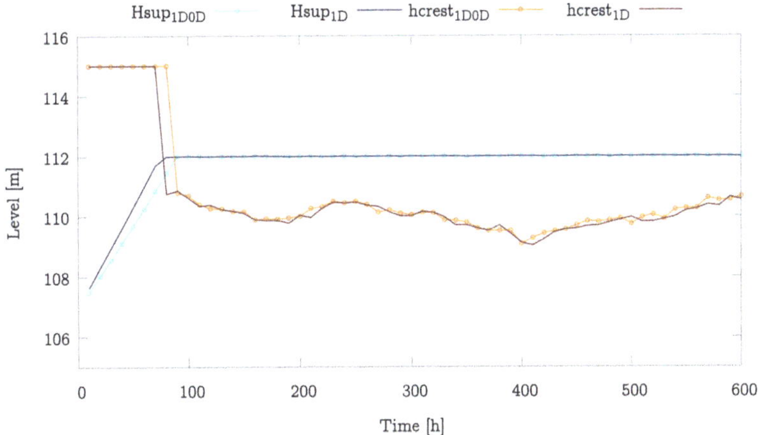

Figure 14. Temporal evolution of the water level and dam crest computed with the fully 1D model and the coupled 1D-0D model for the 2015 case with the control of a PID algorithm.

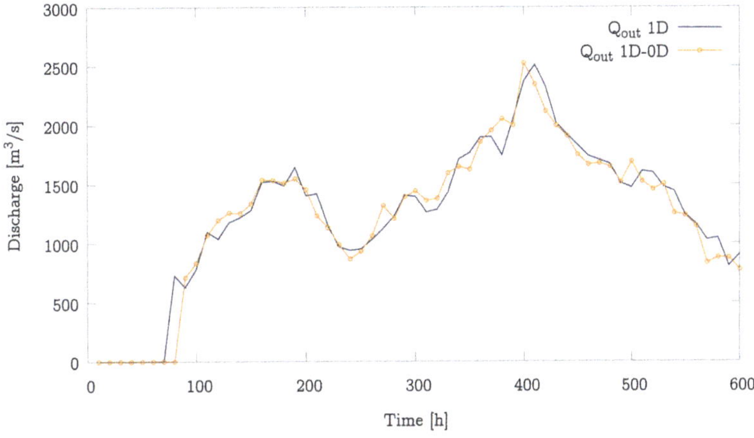

Figure 15. Temporal evolution of the outlet discharge at Mequinenza dam (E003) computed with the 1D model and the coupled 1D-0D model for the 2015 case with the control of a PID algorithm.

The computational times for the model with the PID algorithm are shown in Table 3. The same trend found for the comparison between the different approaches for the reservoir modellization results into an improvement of the optimization when using the 1D-0D model. Besides that, the PID algorithm does not penalize the computational time, but it makes the model more efficient.

Table 3. Computational times for 2015 flood event simulated in the Ebro River with the 1D-0D model and the pure 1D model, both with the dam regulated with a PID algorithm.

Event	Duration	Pure 1D	1D-0D
2015	600 h	484 s	176 s

5. Conclusions

In this work, the performance of several modelling approaches has been compared in order to evaluate their results and computational requirements in a transient river flow event in a reach of the Ebro river (Spain) that includes a reservoir covering a large area. A 2D distributed shallow water model solved over a triangular grid and a 1D shallow water model have been used to discretise the full domain. Additionally, an aggregated volume balance model has been implemented to model the reservoir region in order to allow computational saving. This has led to a coupled 1D-0D model. Finally, a PID control algorithm has been implemented as a regulation technique at the dam location and has been combined with both the 1D model and the 1D-0D model.

From the comparison of the performance of the 2D and 1D models, it can be be concluded that the results of the 1D model for the recent flooding events at the considered Ebro River reach are very similar to those provided by the 2D model. The water level and discharge data predicted by both models follow the same trend. The cross sections used to build the 1D model computational mesh were carefully located to reproduce the river curvature in detail, which is important to obtain a realistic evolution of the hydraulic variables. This effort is justified by the immense computational saving that the use of the 1D model offers, as long as there is no interest in representing the floodplain flow, that the 1D model does not take into account.

The coupling of the 1D model for the river flow at the upstream reach and the 0D model for the reservoir (1D-0D model) offers results very similar to those from the full 1D model. There is some lag due to the instantaneous propagation of the hydrograph in the reservoir assumed by the 0D model but this is acceptable considering the computational savings that the use of this model implies compared to the full 1D model. The computational times observed with the 1D-0D model justifies the use of this combined approach. Therefore, the coupling of a 0D model for the reservoir with the 2D model for the upstream river reach is envisaged as future work since this will lead to high computational savings, something very positive for simulations with 2D models as well as the possibility to simulate the floodplain flow behaviour.

The PID control algorithm has been implemented with the objective to ensure a fixed surface water level at the dam. The results show that this target level value is never reached despite the time variable discharge, which means that the implementation of the control algorithm is a correct security measure to avoid exceeding certain levels in the reservoir. It will be convenient in the future to implement an algorithm that takes into account more realistic and complex objectives.

Author Contributions: Conceptualization, P.G.-N. and I.E.; methodology, P.G.-N. and I.E.; software, I.E.; 2D simulations J.M.; 1D model validation, P.V., I.E. and P.G.-N.; formal analysis, I.E., P.G.-N.; writing original draft preparation, P.V.; writing review and editing, I.E. P.G.-N.; visualization, P.V.; supervision, P.G.-N.; project administration, P.G.-N.; funding acquisition, P.G.-N. All authors have read and agreed to the published version of the manuscript.

Funding: This research was partially funded by the PGC2018-094341-B-I00 research project of the Ministry of Science and Innovation/FEDER. The authors would like to thank also the Confederación Hidrográfica del Ebro staff for their availability and for the supply of the data. Additionally, Isabel Echeverribar was wants to thank to the MINECO for his Research Grant DIN2018-010036.

Institutional Review Board Statement: Not applicable.

Informed Consent Statement: Not applicable.

Data Availability Statement: River measurements are available at http://www.saihebro.com/saihebro/index.php?url=/datos/mapas/tipoestacion:A; accessed on 20 October 2021.

Acknowledgments: The authors acknowledge the CHE for the data availability and their support. The authors also would like to thank all collaborators for their help performing the 2D simulations: Pilar Brufau and Mario Morales.

Conflicts of Interest: The authors declare no conflict of interest. The funders had no role in the design of the study; in the collection, analyses, or interpretation of data; in the writing of the manuscript, or in the decision to publish the results.

References

1. C. o. t. E. U. European Parliament 2007 Directive 2007/60/EC of the European Parliament and of the Council of 23 October 2007 on the assessment and management of flood risks. In *EU Directive*. 2007. Available online: https://eur-lex.europa.eu/legal-content/EN/TXT/?uri=celex:32007L0060 (accessed on 10 October 2021).
2. Centre for Research on the Epidemiology of Disasters (CRED). *The Human Cost of Weather Related Disasters (1995–2015)*; CRED: Brussels, Belgium, 2015; pp. 1–30.
3. Tanoue, M.; Taguchi, R.; Alifu, H.; Hirabayashi, Y. Residual flood damage under intensive adaptation. *Nat. Clim. Chang.* **2021**, *11*, 823–826. [CrossRef]
4. Leandro, J.; Schumann, A.; Pfister, A. A step towards considering the spatial heterogeneity of urban key features in urban hydrology flood modelling. *J. Hydrol.* **2016**, *535*, 356–365. [CrossRef]
5. GebreEgziabher, M.; Demissie, Y. Modeling Urban Flood Inundation and Recession Impacted by Manholes. *Water* **2020**, *12* , 1160. [CrossRef]
6. Vacondio, R.; Aureli, F.; Ferrari, A.; Mignosa, P.; Palù, A. Simulation of the January 2014 flood on the Secchia River using a fast and high-resolution 2D parallel shallow-water numerical scheme. *Nat. Hazards* **2016**, *80*, 1–23. [CrossRef]
7. Echeverribar, I.; Morales-Hernández, M.; Brufau, P.; García-Navarro, P. 2D numerical simulation of unsteady flows for large scale floods prediction in real time. *Adv. Water Resour.* **2019**, *134*, 103444. [CrossRef]
8. Morales-Hernández, M.; Murillo, J.; García-Navarro, P. The formulation of internal boundary conditions in unsteady 2-D shallow water flows: Application to flood regulation. *Water Resour. Res.* **2013**, *49*, 471–487. [CrossRef]
9. Echeverribar, I.; Morales-Hernández, M.; Brufau, P.; García-Navarro, P. Use of internal boundary conditions for levees representation: application to river flood management. *Environ. Fluid Mech.* **2019**, *19*, 1253–1271. [CrossRef]
10. Morales-Hernández, M.; Sharif, MD. B.; Kalyanapu, A.; Ghafoor, S.K.; Dullo, T.T.; Gangrade, S.; Kao, S.-C.; Norman, M.R.; Evans, K.J. TRITON: A Multi-GPU open source 2D hydrodynamic flood model. *Environ. Model. Softw.* **2021**, *141*, 105034. [CrossRef]
11. Chen, J.; Hill, A.A.; Urbano, L.D. A GIS-based model for urban flood inundation. *J. Hydrol.* **2009**, *373*, 184–192. [CrossRef]
12. Ghansah, B.; Nyamekye, C.; Owusu, S.; Agyapong, E. Mapping flood prone and Hazards Areas in rural landscape using landsat images and random forest classification: Case study of Nasia watershed in Ghana. *Civil Environ. Eng.* **2021**, *8*, 1923384.
13. Horritt, M.S.; Bates, P.D. Evaluation of 1D and 2D numerical models for predicting river flood inundation. *J. Hydrol.* **2002**, *268*, 89–99. [CrossRef]
14. Murillo, J.; García-Navarro, P.; Burguete, J.; Brufau, P. The influence of source terms on stability, accuracy and conservation in two-dimensional shallow flow simulation using triangular finite volumes. *Int. J. Numer. Methods Fluids* **2007**, *54*, 543–590. [CrossRef]
15. Kalyanapu, A.J.; Shankar, S.; Pardyjak, E.R.; Judi, D.R.; Burian, S.J. Assessment of GPU computational enhancement to a 2D flood model. *Environ. Model. Softw.* **2011**, *26*, 1009–1016. [CrossRef]
16. Murillo, J.; García-Navarro, P. Energy balance numerical schemes for shallow water equations with discontinuous topography. *J. Comput. Phys.* **2013**, *236*, 119–142. [CrossRef]
17. Sampson, C.C.; Smith, A.M.; Bates, P.D.; Neal1, J.C.; Alfieri, L.; Freer, J.E. A high-resolution global flood hazard model. *Water Resour. Res.* **2015**, *51*, 7358–7381. [CrossRef]
18. Caviedes-Voullième, D.; García-Navarro, P.; Murillo, J. Influence of mesh structure on 2D full shallow water equations and SCS curve number simulation of rainfall/runoff events. *J. Hydrol.* **2012**, *448*, 39–59. [CrossRef]
19. Bomers, A.; Schielen, R.M.J.; Hulscher, S.J.M.H. The influence of grid shape and grid size on hydraulic river modelling performance. *Environ. Fluid Mech.* **2019**, *19*, 1273–1294. [CrossRef]
20. Fread, D.; Hsu, K. Applicability of Two Simplified Flood Routing Methods: Level-Pool and Muskingum-Cunge. In Proceedings of the ASCE National Hydraulic Engineering Conference, San Francisco, CA, USA, 25–30 July 1993; pp. 1564–1568.
21. Nanía, L.S.; Gómez, M. *Ingeniería Hidrológica*; Grupo Editorial Universitario: Granada, Spain, 2004.
22. Haun, S.; Olsen, N.R.B. Three-dimensional numerical modelling of reservoir flushing in a prototype scale. *Int. J. River Basin Manag.* **2012**, *10*, 341–349. [CrossRef]
23. Mohammad, M.E.; Al-Ansari, N.; Issa, I.E.; Knutsson, S. Sediment in Mosul Dam reservoir using the HEC-RAS model. *Lakes Reserv. Res. Manag.* **2016**, *21*, 235–244. [CrossRef]
24. Kawara, O.; Yura, E.; Fujii, S.; Matsumoto, T. A study on the role of hydraulic retention time in eutrophication of the Asahi River Dam reservoir. *Water Sci. Technol.* **1998**, *37*, 245–252. [CrossRef]
25. Bellos, C.; Hrissanthou, V. Numerical simulation of morphological changes in rivers and reservoirs. *Comput. Math. Appl.* **2003**, *45*, 453–467. [CrossRef]
26. Henderson, F.M. Open channel flow. In *Macmillan Series in Civil Engineering*; McGraw-Hill: New York, NY, USA, 1966.
27. Chow, V.T.; Maidment, D.R.; Mays, L.W. *Applied Hydrology*; McGraw-Hill: New York, NY, USA, 1988.
28. Fiorentini, M.; Orlandini, S. Robust numerical solution of the reservoir routing equation. *Adv.Water Resour.* **2013**, *59*, 123–132. [CrossRef]

29. Liu, Y.; Yang, W.; Wang, X. Development of a SWAT extension module to simulate riparian wetland hydrologic processes at a watershed scale. *Hydrol. Process.* **2008**, *22*, 2901–2915. [CrossRef]
30. Dorchies, D.; Thirel, G.; Jay-Allemand, M.; Chauveau, M.; Dehay, F.; Bourgin, P.-Y.; Perrinb, C.; Joste, C.; Rizzolie, J.L.; Demerliac, S.; et al. Climate change impacts on multi-objective reservoir management: Case study on the Seine River basin, France. *Int. J. River Basin Manag.* **2014**, *12*, 265–283. [CrossRef]
31. Cohen Liechti, T.; Matos, J.P.; Ferràs Segura, D.; Boillat, J.-L.; Schleiss, A.J. Hydrological modelling of the Zambezi River Basin taking into account floodplain behaviour by a modified reservoir approach. *Int. J. River Basin Manag.* **2014**, *12*, 29–41. [CrossRef]
32. Ginting, B.M.; Harlan, D.; Taufik, A.; Ginting, H. Optimization of reservoir operation using linear program, case study of Riam Jerawi Reservoir, Indonesia. *Int. J. River Basin Manag.* **2017**, *15*, 187–198. [CrossRef]
33. Arcement, G.; Schneider, V. *Guide for Selecting Manning's Roughness Coefficients for Natural Channels and Flood Plains*; No. 2339. U.S. Geological Survey. Water-Supply Paper. USGS Publications Warehouse; 1984. Available online: https://pubs.er.usgs.gov/publication/wsp2339 (accessed on 10 October 2021).
34. Toro, E.F. The Riemann Solver of Roe. In *Riemann Solvers and Numerical Methods for Fluid Dynamics*; Springer: Berlin/Heidelberg, Germany, 1997.
35. Murillo, J.; García-Navarro, P. Weak solutions for partial differential equations with source terms: Application to the shallow water equations. *J. Comput. Phys.* **2010**, *229*, 4327–4368. [CrossRef]
36. Morales-Hernández, M.; Petaccia, G.; Brufau, P.; García-Navarro, P. Conservative 1D–2D coupled numerical strategies applied to river flooding: The Tiber (Rome), *Appl. Math. Model.* **2016**, *40*, 2087–2105. [CrossRef]
37. Lacasta, A.; Juez, C.; Murillo, J.; García-Navarro, P. An efficient solution for hazardous geophysical flows simulation using GPUs. *Comput. Geosci.* **2015**, *78*, 63–72. [CrossRef]
38. Morales-Hernández, M.; García-Navarro, P.; Burguete, J.; Brufau, P. A conservative strategy to couple 1D and 2D models for shallow water flow simulation. *Comput. Fluids*, **2013**, *81*, 26–44. [CrossRef]
39. Leveque, R. *Numerical Methods for Conservation Laws Lectures in Mathematics*; ETH Zürich; Birkhuser: Zürich, Switzerland; Basel, Switzerland, 1992.
40. Sotelo, G. *Hidráulica General*; Limusa: Johannesburg, South Africa, 2002; Volume 1.
41. Panda, R.C. *Introduction to PID Controllers*; IntechOpen: London, UK, 2012.

MDPI
St. Alban-Anlage 66
4052 Basel
Switzerland
Tel. +41 61 683 77 34
Fax +41 61 302 89 18
www.mdpi.com

Water Editorial Office
E-mail: water@mdpi.com
www.mdpi.com/journal/water

www.ingramcontent.com/pod-product-compliance
Lightning Source LLC
LaVergne TN
LVHW070843170726
843515LV00005B/562

* 9 7 8 3 0 3 6 5 3 3 1 7 9 *